Titles in This Series

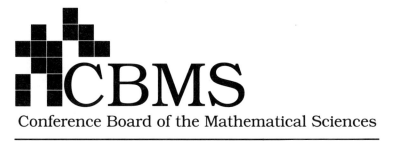

CBMS

Conference Board of the Mathematical Sciences

Issues in Mathematics Education

Volume 4

Research in Collegiate Mathematics Education. I

Ed Dubinsky
Alan H. Schoenfeld
Jim Kaput
Editors

Thomas Dick
Managing Editor

American Mathematical Society
Providence, Rhode Island
in cooperation with
Mathematical Association of America
Washington, D. C.

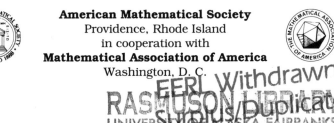

1991 *Mathematics Subject Classification.* Primary 00–XX, 92–XX.

ISBN 0-8218-3504-1

ISSN 1047-398X

CONTENTS

Preface

Welcome to the inaugural issue of *Research in Collegiate Mathematics Education,* part of the *Issues in Mathematics Education* series. The field of research in collegiate mathematics education (RCME) has grown rapidly over the past 25 years, from essentially nil in the 1970's to the point where sessions on educational issues at the annual mathematics meetings now draw large crowds. There is now a body of work, and a growing body of researchers, addressing both basic and applied issues of mathematics education at the college level. It is our pleasure to present some of that work. The introductory articles, survey papers, and exemplars of current research that appear in this volume convey some aspects of the state of the art.

We hope to serve two audiences. *RCME* is for researchers in collegiate mathematics education, and for the wider community of mathematicians who may be interested in these issues both for fundamental intellectual reasons and because of applications to their instruction. In trying to serve these two constituencies, we strive to maintain a delicate balance. On the one hand, there is a minimum technical threshold for scholarly articles in any field. Imagine, for example, what it would be like if the editors of a volume called *Research in Topology* advised potential authors to avoid technical terms such as "compact" or "cohomology" because general readers might not know what they mean! On the other hand, we want articles to be readable and useful beyond the community of researchers — indeed, we want to make the research accessible, and perhaps even lure mathematicians into doing work on education. Thus we have asked authors to keep the general audience in mind. They have been asked to provide general background for the issues they address, to avoid using jargon when plain language will suffice, and to look for and discuss applications where possible. Yet, the general reader must expect to work as well — just as the algebraist has to work to follow a talk on ergodic theory, or the non-specialist has to work to get the gist of the proof of Fermat's last theorem.

As Alan Schoenfeld's opening chapter makes clear, mathematicians who are not familiar with educational research are likely to be in for some surprises. The field is not what you might think it is, and its methods are not what you might expect. This should come as no great shock. Compare people's preconceptions of what a mathematician does with the reality of being a mathematical researcher. The same applies here.

We offer one last point to help calibrate your expectations regarding papers in *RCME*. The default expectation among many teachers of collegiate mathematics is that educational research is all about how to teach some particular course better, or about answering questions such as "Is cooperative learning more effective than whole-class instruction?" While research in collegiate mathematics education is both a pure and an applied field, the applications are not necessarily direct. The main reason for this is that the educational problems to be solved are complex, layered, and subtle; they require large webs of knowledge built by researchers coming from several different directions. For example: There is not just one approach to cooperative learning, but many, which differ in their particulars. The ways they are implemented depend on the characteristics of the instructors, and the ways they are received depend on the (not well understood) characteristics of the students. Some topics may be more amenable to group work than others — and understanding why this is the case requires understanding of the conceptual structure of those topics. Understanding how and why students come to grips with ideas through cooperative learning will require theories of learning and of group dynamics, and ways to measure progress in terms of those theories. In short, answering the cooperative learning question requires integrating knowledge from many sources. The answer is likely to be complex, even though we yearn for simple answers. Despite our most fervent hopes, good educational research is no more likely to provide quick and easy solutions to our educational problems than medical research is to provide quick and easy solutions to our health-care problems. Nonetheless, we can and do expect real progress from both medical and educational research. We hope that this volume and the ones to follow make a significant contribution to that progress.

Let us now introduce the papers.

In "Some notes on the enterprise," Alan Schoenfeld offers an introduction to research in collegiate mathematics education, and the kinds of work that readers might expect to see in this volume and its successors. In the educational analogue of an introductory "what is contemporary research in mathematics about?" expository paper, he describes examples of mainstream work and research methods. Readers will find descriptions of a wide range of studies, ranging from "large n" studies that employ statistical analyses to very small n studies that use "clinical interviews" or analyze videotapes of problem-solving performance.

In "Students, Functions, and the Undergraduate Curriculum," Patrick Thompson provides a selective review of research on students' understanding of functions and its importance for the undergraduate curriculum. This wide-ranging essay deals with a number of basic issues: the nature of conceptual growth, aspects of what it means to understand "function" (e.g., function as action, process, object, or correspondence); issues of representation; student cognition; and instructional obstacles. The article also offers pointers to the broad literature on student understanding of functions.

A more fine-grained study of function, exemplifying the detailed studies from which Thompson draws, is given in Eisenberg and Dreyfus's article "On understanding how students learn to visualize function transformations." Eisenberg and Dreyfus present an analysis of what *function sense* can mean, comparing it with *number sense*. Their emphasis is on visualization. We know that many mathematicians rely heavily on visual approaches to problem situations. It seems that our students do not, and that this contributes to their low levels of success in learning mathematics. Eisenberg and Dreyfus believe that visualization can be taught. Concentrating on transformations of a function's graph resulting from adding and multiplying by constants, they apply the *Green Globs* software in teaching experiments to help students think in terms of translating, stretching, and dilating. Their research methodology combines testing and interviewing. Their results suggest that their methods lead to progress, but that much more needs to be done. They find that students are considerably more likely to be successful with explicit functions that are both simple and familiar to them, but that thinking about properties of general functions remains difficult. Some theoretical reasons for these difficulties are offered.

One subject area that has received a great deal of attention in recent years has been calculus. The four calculus-related articles in this volume indicate the broad diversity of approaches to the topic, and the broad range of research methods currently being employed.

In "Three approaches to undergraduate calculus instruction," Sandra Frid studied the performance of students who had taken three kinds of introductory calculus classes at three different institutions: traditional technique-oriented, "concepts first" (followed by technique), and an approach based on infinitesimals. She identified three different types of students occurring across all forms of instruction: Collectors, Technicians and Connectors. These types differed significantly in the degree to which their sources of conviction were based in their own reasoning vs. external authority. The reader is likely to recognize these types from Frid's descriptions. Students' language use also varied according to the kind of instruction they had received, particularly in the class using the infinitesimal approach. Frid's characterization of the three types of students raises issues that deserve careful follow-up. Might these individual differences, which interplay with different approaches to instruction in important ways, merely reduce to what most people would describe as differences in ability, or do they capture more subtle intellectual/personal differences?

New models of calculus instruction have received significant attention, and are beginning to receive careful research scrutiny as well. Two approaches to calculus that have been widely discussed are the "laboratory-based" approach to calculus developed by Project CALC at Duke, and the "calculus workshop model" developed by Uri Treisman at Berkeley. In "A comparison of the prob-

lem solving performance of students in lab based and traditional calculus," Jack
Bookman and Charles P. Friedman report the results of a preliminary assessment of student understanding in Project CALC. The issues are subtle, in that
the comparison is, in part, one of apples and oranges: students who study the
calculus via computer-based mathematics laboratories and those who work from
a standard textbook have different experiences, and the instruction they receive
can have different (though not contradictory) goals. Finding a common ground
for evaluation, and highlighting "what counts" in the students' experience, is a
challenging task. Bookman and Friedman indicate ways in which such comparisons can be made, and document the results.

Providing adequate instruction for underrepresented groups in collegiate mathematics education (and other fields) is of great national concern. There is considerable interest in Treisman's calculus workshop model for African-American
students. In "An efficacy study of the calculus workshop model," Martin Bonsangue studies its effectiveness with one group of students at California Polytechnic State University, Pomona. In this longitudinal study the academic performance, responses to a questionnaire, and interviews of students who experienced
a version of Treisman's model are compared with those of students who did not
have such experiences. Bonsangue explores three questions. First, he finds that
"African American and Latino students who participated in the workshop calculus sessions achieved as high or higher than any other ethnic group of students,
both during the first year and beyond." Second, there is evidence suggesting
that this achievement was, at least in part, a result of developing student talent
rather than a function of selection. Third, he found some evidence of feelings
of ethnic separation and academic inadequacy among those minority students
interviewed, despite their success. Also, Bonsangue offers some data suggesting
that the per-student cost of the workshop program was much less than the cost
of re-taking courses by students who were not in the program.

Finally, Steve Monk and Ricardo Nemirovsky provide a detailed look at one
aspect of the conceptual underpinnings of calculus. In "The case of Dan" they
examine the nature of a single student's evolving understanding of the relations
between a time-based graph of physical quantities (measures of air flow) and
the actual phenomena, as the student worked with physical apparatus that was
computer-linked to the graphing system. In a close analysis of a video and transcript record of the student's work over a lengthy session, the authors describe
how the student uses, and is confused by, the visual features of the graphs of
air flow as he manipulates the apparatus and reflects on the graphs that result.
Of special interest is the way Dan's conceptions evolved in the face of conflicts
between his interpretations and expectations on one hand, and the phenomena
that he produced on the other. Despite all the messiness of Dan's behavior and
thinking, a significant amount of structure can be discerned, and this structure
turns out to be the means by which we can account for the evolution of his understanding during the session. It is important to realize, however, that this was

not a teaching session, so that the interviewer was not trying to create conceptual change, but rather was trying to probe Dan's understanding, and occasionally, direct his attention to features in the situation that might make more apparent to Dan the consequences of his assumptions or the gaps in his explanations.

Mary Margaret Shoaf-Grubbs' article "The effect of the graphing calculator on female students' spatial visualization skills and level-of-understanding in elementary graphing and algebra concepts" also explores aspects of graphing, but from a very different perspective. Shoaf-Grubbs used a broad battery of tests to examine the spatial visualization and content performance skills of two groups of students, who took alternate versions of a course in college algebra. One group consistently used graphing calculators as part of their instruction, while a second group studied the same content, absent the technology. The data, some of which are presented via a novel use of scattergrams, show significant gains for the group that used the calculators.

Rina Zazkis and Helen Khoury's article "To the right of the 'decimal' point" explores the understandings of pre-service mathematics teachers related to the representation and transformation of numbers in various number bases. The authors attempt to apply the "action-process-object" framework to explain what it means to understand such transformations. Because of its theoretical emphasis, this paper contains a fair number of terms that are likely to be unfamiliar; hence some care and attention are necessary to reap the rewards of reading it. Zazkis and Khoury apply their theories to data generated by interviews of students learning about transforming non-integer numbers from base 5 to base 10. They find, as have others, that preservice teachers' understanding of mathematics is fragile and incomplete — and that although such weak understanding may not interfere with their performing various algorithms correctly, it might well result in less effective instruction for their future students. The authors' explanations of student performance suggest certain generalities regarding individuals' construction of mathematical knowledge — generalities that, they suggest, have potential for application to learning more advanced collegiate mathematics as well as the topics they consider.

This volume concludes with Lynn Steen's "Twenty Questions about Research on Undergraduate Mathematics Education." Through the years, Steen's "20 Questions" papers have focused the mathematical community's attention on important issues regarding mathematical and pedagogical efforts. Here, Steen provides the same service for mathematics educators. His questions provide grounds for reflecting on the contents of this volume, and a challenge for the authors of future articles in the field.

Again, welcome. We hope you both enjoy and learn from what follows.

Ed Dubinsky
Jim Kaput
Alan Schoenfeld

CBMS Issues in Mathematics Education
Volume 4, 1994

Some Notes on the Enterprise
(Research in Collegiate
Mathematics Education, That Is)

ALAN H. SCHOENFELD

OVERVIEW

This essay offers a personal view of recent research in collegiate mathematics education. My goal is to convey the flavor of the work as I understand it–to indicate main themes and to provide some examples that illustrate them. I begin with a context-setting analogy and a brief discussion of recent history that suggests the scope of the enterprise and the variety of research methods currently used. That introduction is followed by a series of examples, each of which illustrates a typical line of research. Broadly speaking, the examples progress from studies that use relatively familiar research methods to studies whose methods are likely to be unfamiliar, from "large n" studies to those where n is very small. The examples are embedded in a running commentary, which elaborates my view of what constitutes productive research.

INTRODUCTION

Just what is research in mathematics education? Like research in mathematics, it is a many splendored thing–and like research in mathematics, it is growing rapidly.

Consider how you would have answered the question "Just what is research in mathematics?" a quarter-century ago, and today. In 1968, the year I entered graduate school, doing mathematics meant proving theorems, by hand. (At least it meant so in pure mathematics, in which I was trained.) It may have been a bit of a stretch, but the standard dictionary definition of mathematics, "the science of space and number," still seemed a pretty good 0th approximation to the mathematical enterprise. That definition seems less adequate in the 1990s.

My thanks to Ming Chiu, Ed Dubinsky, Cathy Kessel, Jeremy Kilpatrick, Anna Sfard, Lynn Steen, and Emily van Zee for their very helpful comments on an earlier version of this paper.

Today, the phrase "mathematical sciences" serves as an umbrella description of the broad body of research in, or steeped in, mathematics. The classical investigative tools of mathematics still lie at the core of the mathematical sciences, but they have been supplemented by a host of others–consider, for example, the use of computers in the proof of the four-color theorem. But it is not just the methods of mathematical inquiry that have expanded in scope. The very nature of mathematics has evolved radically. The most notable example, of course, is that the computer has become both a tool and an object of inquiry: witness the empirical investigations in algebraic geometry made possible only by powerful computational technology, studies of chaos and fractals, complexity theory, and applications of pure mathematics such as CAT scans and public key cryptography. In short, things have changed. On those grounds, Steen argued in 1988 that it might be good to reconceive mathematics as the science of patterns.

There has been a comparable explosion in the scope and methods of research in mathematics education. In 1970–the year that the *Journal for Research in Mathematics Education* published its inaugural volume–the prevailing methodologies were statistical. Researchers in education borrowed many empirical methods from the physical and experimental sciences, including "treatment A versus treatment B" designs and factor analyses. Given the right conditions, such studies could be truly informative. If, for example, two fields of corn were treated almost identically and there was a significant difference in yield between them, that difference could presumably be attributed to the difference in treatment–be it more or less watering, the use of two different fertilizers, etc. Likewise, blind medical tests are based on the same premise. Statistical analyses can suggest whether one treatment–e.g., particular drugs, exercise, or diet–has significantly different effects than another.

The educational analogue went as follows. Researchers created an instructional treatment based on ideas about what contributes to mathematical learning. They then had two groups of students study the same material. The treatment group received the special instruction as designed, and a control group used either a conventional approach to the subject matter or a specially designed method that omitted the key components of the new treatment. Various tests were given to both groups, before and after the instruction. If there were pretest-to-posttest differences in student learning between the two groups, those differences were attributed to the differences in instruction. Another type of study was based on factor analysis. In such work, students were given a battery of tests—tests of spatial perception, computational ability, verbal ability, and so on. Researchers then looked for correlations between performance on these tests with performance on (for example) tests of mathematical problem solving. Factor analyses were performed to determine which abilities accounted for what percentages of the variance in problem solving performance.

The educational research community came increasingly, through the 1970s and 1980s, to see the limitations of this kind of research. First, instructional treatments were often not well defined: "small group learning" or "teaching problem solving strategies" would be interpreted in different ways by different experimenters, and as a result the research findings varied tremendously. It was not at all unusual for a literature review to indicate that half the studies that were reviewed indicated significant positive results for a given treatment, while half produced no results or negative results. But, that was not the only problem. Classes of students, instructional treatments, and teachers, are often not uniform in their properties, in the ways that medical and agricultural experimentation presume them to be. If different teachers taught the two instructional treatments, then a difference in outcomes might be due to the teachers, not the materials. Or, it might be due to what has been called, loosely and somewhat inaccurately, the "Hawthorne effect": the suggestion that the motivational effects of any new treatment encourage participants to work harder and thus do better, independent of the merits of the treatment.[1] If the same teacher used the materials, the difference might still be attributable to Hawthorne effect, or perhaps to differences that had been ignored–e.g. the fact that one class met before lunch and one after, or . . .

In 1978, Kilpatrick noted that the goal of being scientific in its methods had deformed the research enterprise in mathematics education. He suggested that the field might do well to work at developing a better understanding of important phenomena before it tried to treat those phenomena "scientifically," and contrasted the sterility of work done in the United States with the richness of the clearly interpretive work being done in the Soviet Union (e.g.,[11, 13]). Similarly, it was suggested that factor-analytic studies were of limited worth. Most of the "abilities" on which they focused turned out to be almost tautologically defined in terms of ability to perform well on tests (that is, you have a certain amount of "verbal ability" if you get a certain score on a test of verbal ability), and the research produced no real explanations of how these abilities might, in meaningful ways, contribute to competent performance. Toward the end of the 1970s I called for a moratorium on such studies until researchers could describe coherent constructs that were represented by these "abilities," and explain how they contributed to competent performance. I doubt that my call had much effect one way or the other, but the zeitgeist continued to change, and statistical methods came to diminish substantially in importance.

The preceding comments are not meant to suggest that such methods should be abandoned, but rather that they have a particular role and place–and that the field has come to recognize that place. Statistical results on small samples, with the data properly gathered and analyzed, can be suggestive. (Indeed, you will see the appropriate use of such statistics in a number of the articles in this volume: note the ways they are used, for example, in the articles by Bookman and Friedman, and Shoaf-Grubbs.) Large-scale statistical analyses,

---

[1]See pages 163-167 of Brown, [4], for a historically accurate discussion of the Hawthorne effect.

when hypotheses are clear and the conditions for analysis are appropriate, can yield quite solid findings. Again, the medical analogy is appropriate. Studies that correlate obesity or cholesterol levels with the frequency of heart disease have strong implications for health practices in general, although they are not deterministic at the individual level. In medicine, detailed physiological studies complement the statistical ones, and provide an explanation of mechanism–an explanation of what happens, and why. Likewise in the educational arena, statistical studies that examine the college mathematics performance of students who took calculus in high school–studies that compare, for example, the first semester college calculus grades of students who took high school courses that are, or are not, considered "college-equivalent"–can have clear policy implications, although they do not provide information at the individual level. More fine-grained studies of student learning in calculus help to explain how and why students learn, and to determine the advantages and disadvantages of particular instructional practices.

In short, we have the potential to be much more sophisticated about the use of statistical methods today, and we are much less sanguine about naive interpretations of statistically significant results. In education broadly, and in mathematics education in particular, the past twenty-five years have seen the flowering of a host of new methods to supplement (in some cases, supplant) statistical approaches. Many of the new methods are borrowed from, or adapted from, research in other fields–from psychology, artificial intelligence, and anthropology, to mention just three contributing disciplines. All such methods, when properly used, provide *theoretically based, disciplined ways of enhancing our understanding of mathematical thinking, learning, and teaching.* This italicized phrase is my working definition of research in mathematics education.

SOME CASES IN POINT

What follows is an eclectic selection of research methods, each described in brief. My intent here is to be neither comprehensive nor deep, but rather to suggest the scope and character of the enterprise.[2] As the methods are laid out, readers may find that they differ substantially not only from the statistical work described above, but from the readers' expectations of what such research might

---

[2] I have made the conscious decision, throughout this essay, to sacrifice scholarly detail in order to preserve a clean (if not reductive) story line. The brief history given in the introduction is notable for what it omits; see [12] for a comprehensive discussion. What follows is also skeletal, in that I omit mention of most of the theoretical antecedents of current work. Here are three examples. (1) The theoretical stance known as the "constructivist perspective" and the methodologies for clinical interviews investigating children's understandings are largely based on the work of Piaget. (2) Recognition of the importance of social processes in learning, and hypothetical mechanisms by which such learning takes place, are to a significant extent grounded in the work of L. S. Vygotsky. (3) Empirical studies of student work in arithmetic, and theoretical models that underlie them, can be traced to Brownell and Thorndike, among others. Readers who want more detail on such work in general would do well to start with [9].

be like. Those of us who come from mathematics are used to universal quantifiers, or at least a certain kind of precision: *every* continuous real-valued function defined on a compact metric space attains a finite maximum and minimum value; certain theorems are true except on sets of measure zero; and so on. The very character of results in education is different. They are typically more suggestive than definitive–again, with medical research being perhaps a better analogy than research in mathematics. There are universals, and important ones.[3] Just as the "germ theory of disease" locates the causality for physical events in objects not visible to the naked eye, and has practical ramifications (you should wash your hands frequently, especially before handling food!), constructs such as "constructivism" and corresponding research on cognitive structures (also not visible to the naked eye) tell us how ideas are likely to "take" and also have practical ramifications (you should make sure new ideas are grounded in current understandings, or they're not likely to be understood!).

Much current work consists of "small n" studies, where important aspects of understanding are unraveled and explored. A major goal is to provide coherent explanations of complex human behavior in complex settings–e.g., accounts of successful learning, problem solving, and teaching, and what makes them successful. (Or, conversely, to identify, at a level of mechanism, consistent difficulties in such a way that the causes of the difficulties might be addressed.) The art and science of research in mathematics education is aimed at doing such work rigorously and analytically, so that the evidence is not anecdotal–but in the sense described below, reliable and replicable. What constitutes "evidence" is an issue: much of the argument is indirect, or of a type very different from that found in mathematics or the physical sciences. Also, much of the work is suggestive: specific findings point the way to general results, which have yet to be established. Because the phenomena we seek to explain are subtle and elusive, "triangulation"–looking at an issue from multiple perspectives, to make sure you've really got a handle on it–is critically important.

Computer Simulations And Their Entailments. One type of research, which began to flourish in the early 1970s, is computer simulation. Although there is virtually no such work at the college level, I begin with this genre because it offers a particular style of evidence recognizable to mathematicians: the existence proof. It also exemplifies what some people mean by "cognitive structures."

The basic idea behind such work, motivated by early work in artificial intelligence (AI), is that there is some consistency to human intellectual processes

---

[3]The idea of "universals" is controversial. Some reviewers of this article noted that it is impossible to confirm universality in what is essentially an empirical, inductive science–and hence that claims of universality are either assumptions (and should be clearly labeled as such) or hypotheses to be tested. The point is well taken. I note, however, that one could say the same about, say, Newton's laws in physics or the gas laws in chemistry. A "law" in science is not taken to be a statement of absolute and infallible truth, but a theoretical statement whose grounding appears to be exceptionally solid.

–and that such processes can be modeled by computer. This approach, while limited in many ways, is important for both historical and scientific reasons. The history has to do with trends in psychology, and ways in which the field has come, increasingly, to focus on what counts. In mid-century the behaviorists held sway. Their argument was (in caricature, but only slightly) that all human behavior could be explained by stimulus-response chains: We act the way we do because we've been "trained" by the feedback we've gotten in our prior experiences. Humans might be more complex than Pavlov's dogs, but the notions are the same: if we get food soon after the bell rings, we start to salivate when we hear the sound. What works for dogs and laboratory rats (which could be trained to do complex tasks) works as well for humans, argued the behaviorists: we need not invoke the concept of *mind,* which is a quagmire, but simply observe, and discuss, *behaviors.* For behaviorists such as J. B. Watson and B. F. Skinner (who held sway for quite some time), "mind" was anathema; "thinking" was an illusion.

Interestingly, artificial intelligence turned this notion on its head. Its criteria for success were behavioral: Newell and Simon's (1972) "General Problem Solver" [15] managed to solve problems in symbolic logic, do cryptarithmetic problems, and play a passable game of chess. Behaviorists couldn't argue with the output, which was there to see. *But,* the programs were abstractions of regularities of human performance. Researchers watched the moves good problem solvers made, abstracted the (mental) strategies that they believed lay behind them, and then codified those mental strategies as computer programs. In this way mental strategies were validated–by machine! And mind was once again the subject of legitimate inquiry. (Note: though this historical disquisition may seem a digression, it's necessary: some proportion of what appears in the pages of *RCME* will be concerned, directly or indirectly, with the character of mental structures and operations, and readers had best be prepared for it!)

So, computational studies of mind *are* legitimate, but how could they be rigorous? In a number of ways. The weakest, theoretically, is that a running program constitutes an existence proof. Mentalists might theorize forever about "how one thinks," but in the abstract, such discussions might be all too much like discussions about how many angels could fit on the head of a pin. The issue: how could one know whether one's ideas about the nature of thinking actually work? When the phenomena are simple enough to model, computer models help. A running model provides a sufficiency argument, in that "thinking this way" clearly produces results. (Failure is informative too. When a model doesn't run, or doesn't produce what you thought it would, it means you didn't have it right.) Of course, the existence of such models doesn't prove that humans work the same ways the models do–any more than the Wright Brothers' success at Kitty Hawk in 1907 proved that birds have engines or fixed wings–but it does, like the flight at Kitty Hawk, provide no-nonsense empirical evidence that certain theoretical ideas work. The next, and more significant level of evidence comes with attempts to match machine performance with human performance. Does the program have the same kinds of successes, and failures? Does it slip up where humans tend to? And, can the properties of the program be used, in plausible ways, to shed light on human behavior? Here the science, and the rigor, come from serious, long-term, iterative attempts to make the models conform to the things they model. The best research, like the best science, is cumulative: a model is proposed, tested against empirical phenomena, and modified (on theoretically principled grounds!) where it is found wanting. Finally, the strictest criterion of rigor is prediction.

Here I'll give one example, which comes from elementary mathematics. It's one of my favorites, because (a) it provides conclusive evidence of the power of building such models, and of "mentalism" in general, and (b) it helps to point out the inadequacy of a standard pedagogical notion regarding "good teaching."

The domain is simple base 10 arithmetic. Kids spend a lot of time in elementary school learning to perform base 10 addition and subtraction. They have trouble with the procedures, and it takes them a long while to master them. Now, it's been known for the better part of a century that students tend to make certain kinds of mistakes when using the standard algorithms for base 10 arithmetic (e.g., students have trouble implementing the subtraction algorithm when they have to "borrow" from a column that has a zero in it). These errors are analogous to the errors we see students make in algebra (e.g., when students replace the expression $(a^2+b^2)^{1/2}$ by the expression $(a+b)$) or when they neglect to follow through on nested differentiations while performing differentiations using the chain rule. In any of these domains, the standard model of a good teacher is that of someone who has lots of different ways to explain the content–the idea being that if one explanation doesn't sink in, another will.

The bottom line of the research on arithmetic learning is this. After very detailed analyses of student work on addition and subtraction problems, Brown and Burton [5] discovered amazing consistency in student mistakes. Some mistakes were random, of course: Everyone is sloppy on occasion, slips up, or makes mistakes for reasons we can't fathom. But the pattern of mistakes Brown and Burton uncovered was, part of the time, remarkably consistent. They identified 38 basic bugs (consistent errors, named "bugs" because they operate in the same consistent way as the bugs in a flawed computer program) and some 500 concatenations of the basic bugs, which accounted for the vast majority of student errors. Ultimately they were able to craft a short diagnostic test (consisting of sixteen questions) that allowed them to identify sources of student mistakes. The analysis of a student's responses to this brief test allowed them, roughly half the time, to predict the specific incorrect answers that the student would get to a new set of problems–before the student worked the problems! It is important to understand the level of detail in their work. Consider the subtraction problem

$$5023$$
$$\underline{-636}$$

Brown and Burton's diagnostic model enabled them to predict, for example, that (1) students diagnosed as having one particular bug would arrive at the incorrect answer of 4087; (2) students with a related but different bug would arrive at the incorrect answer of 4687; and (3) students with yet a different bug would arrive at the answer 2497. This level of detail far exceeds that of earlier work. (Also, the success rate for predictions goes up markedly with more extended diagnoses.)

The work just described is intended by its authors not only to be predictive, but also to be psychologically plausible. Thus, characterizing the reasons that students make such mistakes is part of the theory. In brief, Brown, Burton, and their colleagues argue that many bugs develop as faulty abstractions of otherwise functional procedures that the students have learned. For example, here is one (vastly oversimplified) story about the origins of the bug that produces answer 3. This bug appears somewhat infrequently and is easily remedied; I choose it simply because its origins are easiest to describe. (For more detail on other bugs and a discussion of implications, see Maurer, [14].)

Typically, students work hundreds of two-column subtraction problems before they move to three-column subtraction problems. They are given the rationale for the subtraction algorithm, but it often goes over their heads. The students focus on the mechanics of the algorithm, and build their understanding of how to execute it from the feedback they get on their work. The following is a procedural description that works for all two-column subtractions: "If you can't do the subtraction in the ones column, borrow one digit from the column to the left and add ten to the column you're working in." It is easy to imagine that this can be mis-abstracted to the following procedure for multi-column subtractions: "If you can't do the subtraction in any column, borrow one digit from the column to the far left and add ten to the column you're working in." This abstracted procedure always gives the correct answer for two-column subtractions, and it works for many three-column subtractions. (Indeed, since problems are usually sequenced so that students encounter "easy" ones first, it may work for quite a few three-column subtraction problems they encounter.) Hence the students may "understand" the procedure (in the same sense that they understood the two-column procedure) and execute it reliably–only to discover that for some inexplicable reason, their answers to some problems are wrong!

There are numerous implications to the kind of modeling in Brown and Burton's work, of which I mention a few. The theory is predictive and thus falsifiable, and as such it meets the standard criteria for research in the physical sciences. This is not the only form of prediction in research in mathematics education, nor is prediction the only kind of criterion by which research should be judged (see below)–but this kind of research represents one basis vector in the research space, and those who are new to the field should find its general form comfortably familiar. Like the work by Newell and Simon cited earlier, it puts yet another nail in the coffin of (extreme) behaviorism–or, to put the case positively, it offers strong support for the idea of modeling mental structures. That is: based on their observations of consistencies in human performance, researchers make inferences about the cognitive structures that produced those consistencies. Those inferences are used to make predictions about future behavior, in this case via computer models. When individuals' subsequent performance is consistent with the predictions, the inferences are supported (in this case, rather well).

The description just given foreshadows the research genre I am about to describe. The general idea is this. On the basis of prior research and/or empirical observations, one makes explicit a set of theoretical assumptions about the nature of mathematical thinking or learning, about cognitive structures, etc.[4] Those assumptions have implications, either about mathematical performance or about learning. As a result, one can craft some form of observations or experiment to look for those implications. The observations may be naturalistic (e.g. one takes notes or makes videotapes in classrooms), or structured, say in the form of interviews. The experiments may be tightly structured, as is the psychologists' wont; or they may be in the form of classroom studies where one hypothesizes that a specific kind of classroom format, or interaction with the material, will be productive–and one then (quite possibly in collaboration with the person responsible for the instruction) takes a very close look at what happens when it is tried. The data–what one sees–are then matched up against the theory. In the case of a good fit, the theoretical notions are substantiated and may be refined; in the case of a not-so-good fit, one may have to go back to the metaphorical drawing board, either theoretically (are the ideas wrong? not adequately specified?) or pragmatically (did the empirical treatment, or the curriculum and the way it was carried out, really reflect the theoretical ideas? If not, how can they be modified, and tried once again?). In the best of all scientific worlds, one proceeds by successive approximation, with refinements of both theoretical notions and empirical techniques. Indeed, the field should proceed that way, with various researchers following up the implications of others' work and putting the

---

[4]There are many as-yet unelaborated assumptions behind the phrase "theoretical assumptions about the nature of mathematical thinking or learning. . . ." My own biases, and I think those of the field as a whole, are that (a) terms have to be well defined, and (b) some sense of explanation or mechanism has to be part of a theoretical stance. An example of an ill-defined statement, and one that would not serve as the basis for solid inquiry, is "students don't learn as much in large classes as they do in small classes." The problem is that both "learning" and "large classes" are ill-defined. What kind of learning, assessed in what ways? And what, besides class size, is varied in the large-versus-small comparison? (For example, students in small recitation sections often get more direct and more extensive feedback on homework and tests. Is this controlled for in a comparison?) Indeed, hypotheses about feedback and its effects begin to get at what I mean by mechanism, and begin to be addressable in rigorous ways.

ideas to the test.

Finally, a brief comment about pedagogy. I mentioned above that the work on arithmetic bugs suggests the inadequacy of a standard pedagogical notion regarding good teaching–namely, that the good teacher has lots of explanations available, so that ultimately one of them will "sink in." Here's why it's inadequate. That conception of teaching assumes, implicitly, that students are recipients of knowledge. It fails to take into account the fact that what the students already understand–whether right or wrong–will shape the ways in which they interpret new information that they encounter. In this case, the (mis)understandings that students have developed–often largely sensible but incorrect abstractions of the procedures they have studied–may render them immune to straightforward explanation, in that what they hear being said may not be what the teacher believes he or she is saying. Hence good explanations, while important, are not enough. Communicating meaningfully with students may require in addition developing an understanding of the ways that students understand things, and finding a way to make connections to those understandings.

Research on "Efficient Resource Utilization" During Problem Solving.

This example and the one in the following section are intended to exemplify the kind of research genre described two paragraphs above–one that involves the progressive refinement of theory and method, and in which the theory is tested from multiple perspectives. The work discussed in this section comes from my research on problem solving. The description runs at breakneck speed– on average, a pace of more than five years per paragraph!–so it is superficial, to say the least. Readers who want more expository substance, along with a broad summary of the literature on problem solving, might want to look at my article "Learning to think mathematically" [**17**]. Those with a healthy thirst for detail might want to look at my 1985 book *Mathematical Problem Solving*.

The research deals with questions of how well people use the knowledge potentially at their disposal, and how the degree of efficiency of their knowledge usage affects their problem solving success or failure. Briefly stated, the claim that has evolved and has been documented over the years is that "knowing the content" isn't enough to ensure success at problem solving; that making poor decisions about how to spend their time and energy while working problems is a significant cause of people's problem-solving failure, even when they know the relevant mathematics. That is: (a) people who go on "mathematical wild goose chases" will expend a good deal of time and energy in the wrong directions, and never get to use what they do know to solve the problems they are trying to solve, and (b) the number of people who do so is large, so the issue is one of significant practical importance.

Over the years that idea has been scrutinized in a number of ways. The first phase of the research, back in the 1970s, consisted of what might be called "plausibility cases through instructional intervention." I provided students with some strategies for using what they did know (for example, a strategy for selecting techniques of integration when the students were confronted with indefinite integrals), and looked to see if their performance improved. It did, and thus the inference that "how you use what you know makes a difference" gained credibility. The second phase of research consisted of analyzing the videotapes of students working on unfamiliar problems. These were good students who "knew" enough mathematics to solve the problems–simple problems in geometry, a max-min problem easier than one for which they had received full credit on the final exam in a multivariate calculus course, and so on. However, they had no clear contextual cues when they worked these problems: e.g., the max-min problem wasn't being worked in the context of a calculus course, and was phrased "What is the largest . . . " rather than "Find the maximum value . . . ." On more than half of the problem-solving attempts the students read the problem, picked a solution path to work on (typically doing so within 30 seconds of having started the problem), and pursued that direction come hell or high water. Often the direction was wrong, and the students were thus doomed to fail in their problem solving attempts. The findings were pretty robust, at least with regard to the spectrum of students I had encountered. Then, the publication of results and methods offered the possibility of wider replication and refinement. Details of the data analysis methods were given in papers describing the work, so that (a) readers could perform the analyses if they wished, and–more importantly, (b) others could replicate and extend the work by having students solve problems of their choosing, and then applying the same analytic methods. Since the publication of the original work there have been numerous replication and extension studies, and the findings have held up. A significant cause of student failure lies in poor resource allocation, and not necessarily in the absence of resources to begin with such as mathematics principles and techniques.

In related work along more positive lines, I performed similar analyses of the attempts of graduates of my problem solving course to solve novel problems. The result was as follows. A significant factor in students' performance at the end of the course was that they managed to curtail wild goose chases, and were thus able to try two or three different approaches of a problem in one problem solving session. With some degree of frequency, the second or third attempt proved successful. This positive evidence reinforces the main thesis: poor "control" causes problem solving failure, and good "control" enables success. That thesis statement serves here as a summary of results–but it can also be taken as a research hypothesis, and subjected to alternate kinds of scrutiny and verification. Indeed, independent evidence providing further reinforcement came in as I was writing this article. The book *Mathematical Problem Solving* provides a description of "expert problem solving," based largely on comparative analyses of problem-solving successes and failures. Such a description gives rise to a natural prediction: one should see a high incidence of these ostensibly productive behaviors in world-class mathematicians, and not nearly as much evidence in good but not equally distinguished mathematicians. DeFranco [7] had an "expert" group of eight mathematicians (who possessed, cumulatively, 12 honorary degrees, numerous prizes and elected positions, and more than 800 publications) and a comparison group of eight other mathematicians (with more than 100 publications but no prizes or elected positions among them) work a collection of non-standard problems. The former group displayed the pattern of behaviors attributed to problem-solving experts in *Mathematical Problem Solving,* while for the most part such behaviors were absent in the latter group. And so, research triangulation continues to refine and test the ideas.

Research On The "Action-Process-Object" Framework.

My second example is the research program followed by Ed Dubinsky and his colleagues [8](see, e.g., [1, 3, 8]). As elaborated in the report by Zazkis and Khoury in this volume, that approach is grounded in particular theoretical notions: an "action-process-object" framework that hypothesizes not only the character of particular mental structures, but the ways in which those structures evolve as people become familiar with a domain. To mention the example that has probably been given the most extensive scrutiny: a function $f : \mathbb{R} \to \mathbb{R}$ can be conceptualized as both a *process* (which, when given an input value x, returns an output value $f(x)$) and as an *object* (whose graph is "picked up whole" and translated vertically one unit by the transformation $f(x) \to f(x) + 1$, or which we think of as an entity when discussing the composition $h = f \circ g$). Competency in the domain–i.e., an ability to use the notion of function in the various ways called for in doing mathematics–requires one to employ both the process and object conceptions of function, and where necessary, to move fluidly from one conception to the other. An extensive body of research, again done in multiple cycles, has documented the role of such conceptions in competent performance, the difficulties students have in developing them (in particular the notion of function as object; see Sfard, [21]), and the use of various kinds of instruction, including the use of computer-based tools, to help students develop the appropriate conceptualizations. The action-process-object framework has been used to explore students' knowledge of induction, their understanding of the concept of group, and so on. It is hypothesized as a general characterization of mental structures, and as such serves as a means of generating hypotheses and predictions about new arenas: the appropriate analyses should, for example, allow for predictions of where students will encounter difficulties in other mathematical domains. For an example of how that idea works out in practice, see the paper by Zazkis and Khoury in this volume.

A Bottom Line Regarding The Nature of the Enterprise.

I chose the examples given above because they show the progressive refinement of ideas through cumulative research, and the multiple ways in which the same phenomenon or hypothesis can be examined and tested. Such cumulative work, involving replication, extension, and triangulation (obtaining convergent evidence from different perspectives) is one major way of overcoming the obvious reliability problems with small n studies. However, I should note here that in this work, as in all such work, "the devil is in the details." In both of the research programs mentioned here (and all others of this type), empirical work is done with "instructional instantiations" of the ideas–instructional programs that, in the minds of the developers, reflect the principles being examined. How much the implementation *really* reflects the underlying theory can always be questioned; likewise, the power and rigor of the underlying theoretical framework to characterize the behavior that is observed should always be subjected to close scrutiny. Such work claims to have a close theoretical-empirical feedback loop, as it should. But remember: astronomy and astrology both claim the same, and I, for one, am less sanguine about the latter than the former. The moral of this tale is that a practicing community of scholars establishes the standards of inquiry, and those standards should be subject to constant scrutiny and debate.

Having said that, let me move to consider some limiting cases of small n: two cases of $n = 1$ and one of $n = 0$ respectively.

A Study of Teaching.

The first case with $n = 1$ introduces research of a somewhat anthropological character. Cooney describes his 1985 study, "A beginning teacher's view of problem solving," as follows:

> This study focused on a mathematics teacher and examined his beliefs about problem solving while he was finishing his preservice training and during his first 3 months of teaching. His beliefs were revealed through interviews, in which he responded to various types of open-ended situations, through a reaction to a report written about his beliefs, and through observations of his teaching and subsequent interviews. Analyses revealed conflicts between his idealism and the reality of classroom practice, as his students were not always receptive to his problem-solving teaching strategy. ([**6**] p. 324)

It is difficult to summarize the results of the study in brief, for the study portrays a complex conflict in beliefs and practice–the conflict between the professed beliefs in problem solving of a beginning teacher, and the ways in which some of his classroom behavior seemed to contradict those beliefs. On the one hand, Fred (the subject of the study) claimed a strong belief in the value of problem solving as a major component of mathematics teaching and learning. In interviews he claimed that the principal activity of mathematics is solving problems and that "if he could be another person when teaching mathematics, he would choose George Pólya because of his performance in the film *Let Us Teach Guessing*" (p. 328). On the other hand, observations of Fred teaching a general mathematics class revealed very little of the spirit of problem solving à la Pólya.

Why? It turns out the answer lies in a number of different causes. First, despite the invocation of Pólya's name, Fred's view of problem solving was, in fact, at some distance from Pólya's. Pólya's view, as Cooney describes it (and I agree) is that problem solving is an integral part of all mathematical activity. In contrast, Fred saw problem solving as a separable part of mathematics: first you learn a bunch of material, and then you can work on some interesting problems with the mathematics you've learned. Indeed, the problems were not necessarily the vehicles for learning, or even doing, good mathematics. As Fred saw them, they were largely *motivational* in nature: an intriguing problem was intended to get the students engaged, and the fact that its solution was purely routine was of no import. Fred saw the use of problems (read: recreational problems) as being largely inspirational, and he was not fundamentally concerned with the mathematics that might be learned from working those (or other) problems.

To compound the issue, Fred faced the normal pressures of a beginning teacher. With multiple classes to prepare and only so much time to prepare them, he relied heavily on textbooks to structure his lessons. (And anyone who has looked at the standard texts sees that "problem solving" was there in name only, if at all.) While students in some of his more advanced classes enjoyed his attempts to liven up their classes with supplemental problems, students in his general math classes did not. In fact, they resented his attempts. They knew that math class consists of practicing the basics, and that's what a math teacher should have them doing. Fred's excursions into problem solving made them uncomfortable–one student said "We reject it as kind of a culture shock." Not surprisingly, a young teacher who is not sure of himself and who faces consistent hostility when he tries activities that he thinks of as largely motivational (despite his assertions of the centrality of problem solving) is likely to abandon the activities that generate the hostility. Hence Fred could say one thing about mathematics instruction, and behave differently in the classroom.

Now, let us step back and consider this as an exemplar of research. As noted above, this is a study with $n = 1$. What in it could be reliable, valid, or universal? Indeed, do such terms even have potential meaning in this context? Well, for starters, how about "insightful" as a term to consider? This is a case study of a teacher saying one thing, and doing another–with very serious consequences for that teacher's students. It is a pretty safe bet that Fred is not unique. Whatever the rhetoric about "problem solving" may be, one can be sure there are many slips, nationwide, between such rhetoric and the reality of classroom instruction. While the details might differ substantially from teacher to teacher, elaborating both Fred's understanding of the mathematics and the main forces that drove his classroom behavior is, at minimum, likely to point to places where other teachers face similar pressures or difficulties. Let me put the preceding statement in pseudo-mathematical language. What Cooney's paper suggested was that teachers' beliefs and actions, which may have been conceived as a function of a small number of variables, were indeed functions of many more variables. The paper suggested what those variables were, and what their contributions were. The stories told in the paper cannot be dismissed as mere anecdotes, because of the way they are tied together–with multiple sources of evidence on the same themes documenting the solidity of the results.

Universality (or at least generality) is another question. At the most general level, the paper points out, as have numerous others, that people's actions may be at significant variance with their professed beliefs; and that unarticulated and perhaps unrecognized beliefs may play a strong role in shaping people's behavior. (For example, students' beliefs about the nature of mathematics can result in their behaving in decidedly anti-mathematical ways: see [17] for detail.) Hence, to develop a comprehensive picture of people's understandings, one should look for consistency in their actions, and for possible underlying causes of that consistency. In the large, this paper can be situated within that context. In a more narrow context–looking at teacher beliefs–this paper, which was one of the first on the topic, served in some ways as a trial balloon. Providing a good description of what explained Fred's behavior proves nothing about anyone other than Fred, of course. But, it does serve to raise a set of issues. The assumption made above is that what affected Fred will affect lots of other teachers. That's a research hypothesis, to be followed up in other work. If it turns out that the field is rich, a large body of studies will elaborate the work begun in early, suggestive studies such as this one. If not, the work ultimately closes in on itself. In this case, the field is alive and well: see [23] for a review.[5] More generally, the mode of analysis used by Cooney–looking to social and anthropological explanations of why people do what they do in complex educational settings–is getting increased use in mathematics education. As might be expected, the methods and the findings are evolving: in the spirit of contemporary anthropological work, for example, some teachers and researchers are writing their stories together.

A Study of Learning.

The second single-subject study I shall discuss is of a radically different character. A few years ago my research group collected seven hours of videotape of a student working on our graphing software in the company of a research group member. Jack Smith, Abraham Arcavi, and I then spent a year and a half analyzing those seven hours of videotape, with the following simple goals: to be able to characterize everything about linear functions that the student knew when she entered our lab, and to chart the growth of her knowledge over the seven hour period in which she interacted with the researcher and the software. The result of the analysis was a 120-page manuscript [19] that posits a particular "knowledge architecture" (a structure for representing the ways people's knowledge is organized) and provides a detailed description of this student's understanding, and the ways it changed, in those terms. For example, misconceptions that were tied to complex knowledge structures could not easily be "fixed": whether right or

---

[5]I note that although the mechanism of genesis is somewhat different, the "trial balloon" phenomenon just described has a precise analogue in mathematics. Often the publication of a new theorem or the occasion of a conference on X may stimulate interest in X. X will gather some attention, and then, based on its ultimate fertility as a research area, X will either thrive or die. Consider a few C's as examples. Catastrophe theory, for all its publicity a few decades ago, seems to have faded. Complexity theory seems to be going strong. Chaos is young, and the jury is out.

wrong, ideas well tied to other ideas tend to be robust and to resist modification. Conversely, ideas not well tied to others can be "fragile"–that is, easily forgotten or mis-remembered. As explained briefly below, our expectation is that these findings (and many others in the paper) are general, and that they have strong implications for instruction.

As in the case of Cooney's study, one can ask a host of questions about reliability, validity, and generalizability. What could possibly be "universal" about one student's learning? The answer, in short: its *character*. In the analysis we claimed that the way our student's knowledge was organized had implications for the ways she would learn new things (or unlearn old ones), and we traced those implications through the data. Moreover, we claim that the organization of this student's knowledge is what is general: that all people put things together in the way this student did, and that in complex domains, all people's "learning trajectories"–the ways in which their knowledge grows and changes–will be the same. This very strong assertion is, of course, a hypothesis to be tested in future research. In the starkest terms, our claim is that if you go out and watch someone engaged in learning a complex subject matter domain, and then perform an analysis of the type that we did, you will see the same patterns of growth and change of mental structures that we described for this student.

A medical analogy may help clarify what I mean. An early paper on normal physiological functioning–say one describing the circulatory system–might well have been based on the detailed analysis of one or two cases. However, the descriptions in that paper might be intended as characterizations of a general structure. A statement to the effect that "if you're a human being functioning independently of any support systems, then you have a circulatory system that is essentially closed, and a heart that serves as a pump driving blood through the system" could, thus, be seen as a universal claim. Our arguments are of a similar character. (But, recall the caveats about "universality" in footnote 3.)

Next, what could possibly be rigorous about such a study? Of necessity, doesn't a study with $n = 1$ have to be anecdotal? The answer is that, once again, the substance is in the details. There can and must be *internal* consistency to such a study–that is, there have to be ways of assuring that judgments about what a student knows are well grounded and defensible. In this case, saying "the student knows X at time T" meant making the following commitment on the part of the analysts: unless there is subsequent evidence that a change in the student's knowledge of X has taken place, then the analysts are compelled to assume that the student knows X for all times $t > T$. Hence all analyses force predictions of future behavior; subsequent explanations could not be ad hoc, because much of the student's knowledge structure had already been posited. Likewise, an ad hoc ingredient in an explanation of a particular episode would trip us up later, since we would then be constrained to live with that ingredient for the rest of the analyses.

(To pursue the medical analogy suggested above, an early description of the circulatory system might be derived from one case. One's descriptions are not necessarily unconstrained because $n = 1$. Here, the constraints of internal consistency required for an adequate description–e.g., the fact that the amount of blood pumped out of the heart should correspond closely to the amount of blood that returns to it!–prevent the introduction of too many random features in the explanation. On the basis of one carefully worked out description, confirming evidence in a few other cases, and consistency with related knowledge, one might be in a good position to make a general claim. The same is the case for our study.)

Finally, what if we're just good at making up stories? In addition to being consistent, judgments in studies such as the one discussed here have to be reliable. That is, others trained in the same analytic methods must see the same things, with great consistency. For a discussion of reliability and validity in cases such as this, see [18]). The issues are serious: as research moves into uncharted territory such as videotape analyses like the one mentioned here, methodological concerns will be of very high priority. The benefits of being exploratory in our research studies and methods are that we are now paying real attention to "what counts." The costs are that we do not have a well developed canon of research methods and tools. The question "How can we trust what's being said, or the methods used to justify it?" is one we will need to ask frequently, as we work to develop such a canon.

Theoretical Studies.

My last example deals with a case where $n = 0$ — that is, a theoretical piece. Each of my previous examples has dealt with empirical work, in which theoretical statements about thinking and learning are put to the test, or at least illustrated by, data of various types (experimental, observational, etc.). However, as suggested in the previous paragraph, there must also be room for pieces that ask fundamental questions about the enterprise–studies that help us to understand what we are doing when we are working in "empirical mode." As a case in point, consider the issue of *representations* as addressed by Kaput [10]. What does it mean, for example, to say that a computer simulation program "represents" someone's mathematical behavior? I won't try to sort out the answer, but I will point to some of the complexities involved in sorting it out. See Kaput's article for details.

For starters, consider the marks made on paper by the person who is being "simulated" by the computer program. That person scrawls down some symbols, which represent various mathematical ideas. As anyone who's studied foundations knows, there's a lot more there than meets the eye: the symbol "5," representing the number "five," can be seen as representing the "successor" (in the sense of Peano) to the symbol "4," and as representing the cardinality of the set of fingers on my left hand, or any set that can be put into one-to-one correspondence with them; or . . . And that's just the relationship between the symbol and the mathematical notion it represents. Now, what about the person doing the symbolizing? That person has a brain, which is doing its own interpreting and symbolizing: somewhere in the tangle of brain cells in the person's head is a representation of the concept, and the ability to recognize the written symbol and have it trigger some of the relevant mental associations. That, of course, is just the human side–the thinker being simulated or modeled. On the machine side, there are (or should be) analogues of all this. Hence one needs to conceive of the correspondences between various parts of the human-mathematics-written symbol systems being simulated and the computer system doing the simulation. But wait! Who is the "one" that serves as the subject of the previous sentence? Perhaps, to begin with, the person programming the simulation. But then the program itself is a representation of the programmer's understanding of the correspondence between the computer system and the human it represents . . . In short, the issues are complicated, and glossing over them can lead to serious problems. Foundational studies are critical.

CONCLUDING COMMENTS

I have now wound my way down from large n to null. As noted above (and as the title of this article suggests), this cursory survey is both personal and selective. On the one hand, I propose that my working definition of research in mathematics education, *theoretically based, disciplined ways of enhancing our understanding of mathematical thinking, learning, and teaching,* might be of some general use. On the other hand, I harbor no illusions about having done full justice to the broad spectrum of work that falls under that rubric. (For example, some of the articles in this volume lie outside the convex hull of the methods I have mentioned.) Readers who want alternative samplings of ideas and exemplars at the college level might want to read [2] (in press) and Selden and Selden [20], and the sources they point to. Readers who want an introduction to the broad spectrum of research in mathematics education should look at the *Handbook of Research On Mathematics Teaching And Learning* [9]. In this brief article I have merely given my biased view of "the way things are." There are sure to be alternate perspectives–different views on the history of the field, on the nature of evidence, on what constitutes a compelling argument, on the character of research programs and directions, and much much more. I hope and expect that the pages of this series will be a comfortable home for discussions of these issues.

REFERENCES

1. Ayers, T., Davis, G., Dubinsky, E., & Lewin, P., *Computer experiences in learning composition of functions*, Journal for Research in Mathematics Education **19(3)** (1988), 246–259.
2. Becker, J. R., & Pence, B. (in press), *The teaching and learning of college mathematics: Current status and future directions*, J. Kaput and E. Dubinsky (Eds.), Selected papers from the San Francisco AMS/MAA special session on research in undergraduate mathematics education, Mathematical Association of America, Washington, DC.
3. Breidenbach, D., Dubinsky E., Hawks, J., & Nichols, D., *Development of the process concept of function*, Educational Studies in Mathematics (1991), 247–285.

4. Brown, A., *Design experiments: Theoretical and methodological challenges in creating complex interventions in classroom settings*, Journal of the Learning Sciences **2(2)** (1992), 141–178.

5. Brown, J. S. & Burton, R. R., *Diagnostic models for procedural bugs in basic mathematical skills*, Cognitive Science **2** (1978), 155–192.

6. Cooney, T., *A beginning teacher's view of problem solving*, Journal for Research in Mathematics Education **16(5)** (1985), 324–336.

7. DeFranco, T., *A perspective on mathematical problem solving based on the performances of Ph.D. mathematicians. Manuscript submitted for publication* (1993), College of Education, University of Connecticut, Storrs, CT 06269.

8. Dubinsky, E. (in press), *A theory and practice of learning college mathematics.*, A. Schoenfeld (Ed.), Mathematical thinking and problem solving, Erlbaum, Hillsdale, NJ.

9. Grouws, D. (Ed.), *Handbook of research on mathematics teaching and learning*, MacMillan, New York, 1992.

10. Kaput, J., *Notations and representations as mediators of constructive processes.*, E. von Glasersfeld (Ed.), Radical constructivism in mathematics education., Kluwer, Dordrecht, Netherlands, 1991.

11. Kilpatrick, J., & Wirszup, I. (Eds.), Soviet studies in the psychology of teaching and learning mathematics. **1-14** (1972), School Mathematics Study Group, Stanford, CA.

12. Kilpatrick, J., *Variables and methodologies in research on problem solving*, L. Hatfield (Ed.), Mathematical problem solving, ERIC, Columbus, OH, 1978, pp. 7–20.

13. Krutetskii, V. A., *The psychology of mathematical abilities in schoolchildren (J. Teller, Trans.).*, University of Chicago Press. (Original work published 1968), Chicago, 1976.

14. Maurer, S., *New knowledge about errors and new views about learners: What they mean to educators and more educators would like to know.*, A. H. Schoenfeld (Ed.), Cognitive science and mathematics education, Erlbaum, Hillsdale, NJ, 1987, pp. 165–188.

15. Newell, A., & Simon, H., *Human problem solving*, Prentice-Hall, Englewood Cliffs, NJ, 1972.

16. Schoenfeld, A. H., *Mathematical problem solving*, Academic Press, Orlando, FL, 1985.

17. Schoenfeld, A. H., *Learning to think mathematically: Problem solving, metacognition, and sense-making in mathematics*, D. Grouws (Ed.), Handbook for research on mathematics teaching and learning, MacMillan, New York, 1992, pp. 334–370.

18. Schoenfeld, Alan H., *On paradigms and methods: What do you do when the ones you know don't do what you want them to? Issues in the analysis of data in the form of videotapes*, Journal of the Learning Sciences **2(2)** (1992), 179–214.

19. Schoenfeld, A. H., Smith, J. P., & Arcavi, A.. A., *Learning*, R. Glaser (Ed.), Advances in instructional psychology, vol. 4, Erlbaum, Hillsdale, NJ, 1993, pp. 55–175.

20. Selden, A., & Selden, J., *Collegiate mathematics education research: What would that be like?*, College Mathematics Journal **24(5)** (1993), 431–445.

21. Sfard, A., *On the dual nature of mathematical conceptions: Reflections on processes and objects as different sides of the same coin*, Educational Studies in Mathematics **22** (1991), 1–36.

22. Steen, L., *The science of patterns*, Science **240** (1988), 611–616.

23. Thompson, A., *Teachers' beliefs and conceptions: A synthesis of the research*, Grouws (Ed.), Handbook of research on mathematics teaching and learning, MacMillan, New York, 1992, pp. 127–146.

UNIVERSITY OF CALIFORNIA, BERKELEY

CBMS Issues in Mathematics Education
Volume 4, 1994

Students, Functions, and the Undergraduate Curriculum

PATRICK W. THOMPSON

Introduction

Someone, I cannot remember who, paraphrased Winston Churchill by saying that mathematics and mathematics education are two disciplines separated by a common subject. The mathematician is primarily concerned with doing mathematics at a high level of abstraction. The mathematics educator is primarily concerned with what it is that one does when doing mathematics and what kinds of experiences are propitious for a person's later successes. Until recently mathematics education research has focused predominantly on the learning and teaching of early mathematics in the school curriculum, so it is natural that practicing mathematicians have found it difficult to relate to mathematics education research. I suspect that the current interest in calculus reform [**21, 63**] and the broader rethinking of the undergraduate curriculum, together with the advent of the AMS/MAA Joint Committee on Research in Undergraduate Mathematics Education, will lead to a wider recognition that mathematics and mathematics education are fundamentally dependent upon one another.

My purpose in writing this paper is to discuss research on students' understanding of fun ctions and its importance for the undergraduate curriculum. Much has been written recently about concepts of function that goes into far greater detail than I will (see [**33, 46, 57, 64**] for extensive reviews). I will impose a somewhat idiosyncratic structure upon this literature to present an overview of research on concepts of function and to highlight issues I believe need greater consideration than they have so far received.

Revised version of a paper presented at the Annual Joint Meeting of the American Mathematical Society and the Mathematical Association of America, San Antonio, 12-16 January, 1993. Preparation of this paper was supported by National Science Foundation Grants No. MDR 89-50311 and 90-96275. Any conclusions or recommendations stated here are those of the author and do not necessarily reflect official positions of NSF.

As a matter of background, I should say a few words about the perspective I bring to this task. We cannot speak strictly about the development of a single concept, such as function. If we have learned anything in mathematics education research it is that a person's thinking does not respect topical boundaries. When analyzing students' concepts of function, we need to keep in mind that the imagery and understandings evoked in students by our probing is going to be textured by their pre-understandings of such things as expressions, variables, arithmetic operations, and quantity. We also need to keep in mind that their mathematical learning has, for the most part, happened in schools, which means that our interpretations of students' performance must be conditioned by our knowledge that they are taught by teachers with their own images of what constitutes mathematics, and that both the learning and teaching of mathematics are conditioned by the cultures (school, ethnic, and national) in which they occur [**5, 12, 56, 77**]. I have tried to capture these background relationships in Figure 1. While ignoring this issue might simplify matters enormously for us as teachers and mathematics education researchers, we do so at the peril of losing and validity of our interpretations and conclusions.

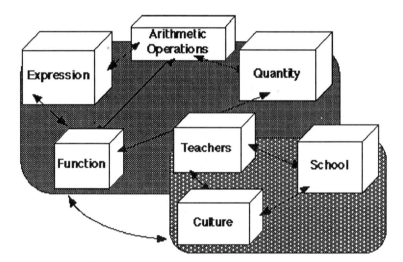

FIGURE 1.   Students' concepts always emerge in relation to other concepts they hold, and in relation to teachers' orientations and in relation to cultural values expressed by teachers, peers, and families.

Another perspective I bring is that concepts emerge over time, much as a dynamical system. The actual form a concept takes (or that it fails to take) in a student's reasoning can be tremendously influenced by seemingly trivial deviations from valid understandings of mathematics they learned much earlier. For example, it is common that elementary school students have impoverished understandings of whole-number numeration. This in itself makes it difficult for them to develop what is often called "number sense"[32], and also makes it virtually impossible for them to make sense of standard arithmetic algorithms. This contributes substantially to their developing an orientation toward memorizing meaningless symbol manipulation. They develop this orientation as a mechanism for coping with an otherwise intolerable situation– not having a clue as to what the teacher is talking about, but nevertheless being expected to perform.

School students' common orientation toward "remembering what to do with marks on paper" eventually shows up in our college classrooms, perhaps showing itself only vestigially as ungrounded formal reasoning. This is what Sfard [59] refers to as disconnected reification–students turning what are offered, by us, as representations into the actual objects of their reasoning. I mention Sfard's notion of disconnected reification for a purpose. Tinkering with instruction or curriculum to emphasize functions will be insufficient if we fail to address students' common orientation to ungrounded symbol manipulation. I will return to this point later.

I will shape my discussion of research on function concepts around six themes. These are:

- Concept image and concept definition
- Function as action, as process, and as object
- Function as covariation of quantities and function as correspondence
- Understanding phenomena and representing phenomena
- Operations on numbers and operations on functions
- Emergent issues

One theme I will not discuss, except to explain why not, is the literature on multiple representations. In this regard I will express my opinion as to why this line of research needs to be rethought.

I selected these six themes because of their emergence in the literature as constructs around which a stable consensus seems to have developed regarding their importance for students' understandings of function. When examined closely, these themes are highly related, but they nevertheless seem to provide a useful organization for entry into the issues of learning and teaching the concept of function and using the concept of function as an organizing construct in the curriculum.

The distinction between students' concept images and the notion of concept definition arose as a way to understand how students expressed reasoning that was inconsistent with taught definitions of function, limit, derivative, etc. The distinction between function as process and function as object emerges from a variety of traditions, both philosophical and psychological. One way to think of this distinction is to reflect on the the formulation $\int_a^x f(t)\, dt$ in the First Fundamental Theorem of Calculus. We must think of integration as the culmination of a limiting process, but at the same time consider that process, applied over an interval of variable length, as producing a correspondence. The third theme,

regarding covariation and correspondence, highlights a tension, both in students'
learning and among researchers, regarding what a function is. I'll attempt to
make it evident that this tension can be both natural and productive. The
fourth theme, regarding phenomena and representation, is one over which I will
linger. It has to do with students' conceptions of the stuff that mathematics
is often about, and I don't just mean physics or chemistry or some domain of
application. The fifth theme, having to do with operations on numbers and operations on functions, reflects my feeling that a distinction needs to become part of
the fabric of instruction in secondary and early undergraduate mathematics–the
distinction between seeing arithmetic operations as operations on numbers and
seeing those same operations as operations on functions. The last theme points
to issues that emerge from research on function concepts that have yet to be
investigated directly.

Concept Image and Concept Definition

The distinction between concept image and concept definition arose originally
in the work of Vinner, Tall, and Dreyfus [66, 83, 85]. In their usage, a concept
definition is a customary or conventional linguistic formulation that demarcates
the boundaries of a word's or phrase's application. On the other hand, a concept
image comprises the visual representations, mental pictures, experiences and
impressions evoked by the concept name.

In lay situations, people understand words through the imagery evoked when
they hear them. They operate from the basis of imagery, not from the basis of
conventional constraints adopted by a community. People understand a word
technically through the logical relationships evoked by the word. They operate
from the basis of conventional and formal constraints entailed within their understanding of the system within which the technical term occurs. Vinner, Tall,
and Dreyfus arrived at the distinction between concept image and concept definition after puzzling over students' misuse and misapplication of mathematical
terms like function, limit, tangent, and derivative. For example, if in a student's
mathematical experience the word "tangent" has been used only to describe a
tangent to a circle, then it is quite reasonable for him to incorporate into his
image of tangents the characteristic that the entire line lies to one side or the
other of the curve, and that it intersects the curve only once [83]. Notice that
this image of tangent–uniquely touching at one point–has nothing to do with
the notion of a limit of secants. It is natural that a student who maintains this
image of tangent is perplexed when trying to imagine a tangent to the graph of
$f(x) = x^3$ at $(0,0)$, or a tangent to the graph of $g(x) = x$ at any point on its
graph.

A predominant image evoked in students by the word "function" is of two
written expressions separated by an equal sign (Figure 2). We might think that
only neophytes hold this image of function. I suspect it is far more prevalent
than we acknowledge. An example will illustrate my suspicion and at the same

time illustrate how Tall, Vinner, and Dreyfus envision the influence of concept images over concept definitions.

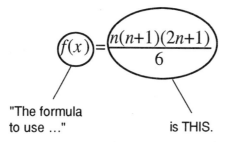

FIGURE 2. A concept image of "function." Something written on the left is "equal to" something written on the right.

My wife, Alba Thompson, teaches a course designed to be a transition from lower- to upper-division undergraduate mathematics. It focuses on problem solving and proof. Students are supposed to take it after calculus and linear algebra, but a fair portion of the class typically have taken at least one term of advanced calculus or modern algebra. In the context of studying mathematical induction she asked one student to put his work on the board in regard to deriving and proving a formula for the sum $S_n = 1^2 + 2^2 + \cdots + n^2$. The student wrote $f(x) = \frac{n(n+1)(2n+1)}{6}$. Not a single student thought there was anything wrong with this formulation. It turned out, after prolonged probing by Alba, that this formulation was acceptable because it fit within students' concept image of function, which I've presented in Figure 3.

$$f(x) = \frac{n(n+1)(2n+1)}{6}$$

"The formula to use ..."   is THIS.

FIGURE 3. Students' acceptance of an ill-formed function representation because it fit their concept image of a function.

An important point to draw from Figure 3 is that malformed concept images are insidious. They keep showing up in the strangest places. On the other hand, we could not function intellectually without having concept images. The key point is that mathematical "experts" come to use concept images and concept definitions dialectically [83, 84]. Over time, their images become tuned so that they are consonant with a conventionally accepted concept definition, which in turn allows intuition to guide and support reason. Not every student of mathematics attains equilibrium between definitions and images, however. We can increase their chances of success by giving explicit attention to imagery as an important aspect of pedagogy and curriculum. In the next sections I discuss important aspects of well-formed concept images of function.

Function as Action, Process, and Object

It is well known that elementary school students have difficulty conceiving of arithmetical expressions as anything beyond a command to calculate [**13, 19, 35**]. They typically do not think of, say, $4(12-(4+5))$ as representing a number. Similarly, algebra students often think of, say, $x(12 - (x + 5))$ as representing a command to calculate. When they come to think of an expression as producing a result of calculating, they have what several researchers have called an *action* conception of function [**6, 24**]. This conception is as a recipe to apply to numbers. Students holding an action concept of function imagine that the recipe remains the same across numbers, but that they must actually apply it to some number before the recipe will produce anything. They do not necessarily view the recipe as representing a result of its application.

When students build an image of "self-evaluating" expressions they have what is called a *process* conception of function. They do not feel compelled to imagine actually evaluating an expression in order to think of the result of its evaluation. From the perspective of students with a process conception of function, an expression stands for what you would get by evaluating it.

It is surprising that achieving a process conception of function is a non-trivial achievement for students, and that for many students it is not achieved without receiving instruction that focuses explicitly on its development [**24, 30**]. Dubinsky and his colleagues [**6, 25**] have developed an instructional approach using ISETL, a set-theoretic programming language, that shows promise as an instructional environment for students' development of a process conception of function. They use ISETL to write named processes, which then serve as function definitions. Students can then direct that a function be applied to individual numbers or to numbers in a pre-specified set, in both cases by using the name of the function in place of its defining process.

A process conception of function opens the door to a wealth of imagery. Students can begin to imagine "running through" a continuum of numbers, letting an expression evaluate itself (very rapidly!) at each number.[1] I should note that to become skilled at conjuring such an image students must practice conjuring it [**18**]. Goldenberg and Lewis [**29, 30**] have developed visual supports for students to envision functions as processes applied over a continuum.

Once students are adept at imagining expressions being evaluated continually as they "run rapidly" over a continuum, the groundwork has been laid for them to

---

[1]I am not speaking literally. Rather, I am speaking from the point of view of the student.

reflect on a *set* of possible inputs in relation to the *set* of corresponding outputs. I will say more about this idea in the section *Covariation and Correspondence.*

At the point where students have solidified a process conception of function so that a representation of the process is sufficient to support their reasoning about it, they can begin (I emphasize begin) to reason formally about functions–they can reason about functions as if they were objects. To reason formally about functions seems to entail a scheme of conceptual operations which grow from a great deal of reflection on functional processes. Primary among these is an image of functional process as *defining* a correspondence between two sets: a set of possible inputs to the process and a set of possible outputs from the process.

The many paths by which students achieve an object conception of function are long and complex [2], and explanations of it draw on a long tradition in philosophy and epistemology regarding the notion of reflective abstraction [4, 23, 53, 54, 55, 69, 86]. One hallmark of a student's object conception of functions is her ability to reason about operations on sets of functions. I should quickly point out that it is easy to be fooled–to think that students are reasoning about functions as objects when it is actually the function's literal representation (i.e., marks on paper) that is the object of their reasoning [58, 59, 61]. I suspect that the kinds of intellectual operations that go into operating meaningfully on functions have a considerable overlap with the kinds of operations that enable students to reason about such things as operations on cosets in quotient groups– behind their visible operations is a tacit image of completed element-by-element operations.[2]

A question raised by several reviewers, and also raised by Carolyn Kieran ([43], p. 232), is whether students must first develop process conceptions of function before developing object conceptions of function. This question was raised in the context of discussions of computer environments that ostensibly allow students to interact directly with function graphs, or to manipulate situations and see real-time changes in associated graphs. My remarks in the section *Multiple Representations* are pertinent to the matter of multi-representational computer environments. For now I will say that only that, as a matter of word usage, I would prefer not to talk about students interacting with functions as objects until I am assured that the *students* have conceived the objects they interact with as functions or representations of functions. As Jim Kaput said, "What is being represented, for a knowledgeable third party observer, is NOT what is being represented for the person living in the representational process" [personal communication]. I think we easily confuse perspectives when we say that students interact with functions as objects. A more veridical description might be that students interact with automatically generated "things" (e.g., wavy

---

[2]Intuitionists might complain that such an image is impossible over infinite domains [7, 36, 80]. But I do not mean that people imagine actually completing all possible element-by-element operations. Rather, they just imagine that it is done—in the same way that they might imagine passing over all possible points between their easy chair and their television set.

lines on a computer screen) that they come to make sense of in relation to the situations the things are tied to and in relation to their progressive internalization of the conventions by which the things behave. So, to answer the question of process/object precedence, I see every reason to believe that in an individual student's construction of function, process conceptions of function will precede object conceptions of function. What has changed because of technological advances are the kinds of experiences we can engender in the hope that students eventually create functions as objects.

Function as Covariation and Function as Correspondence

One way to think of the evolution of today's many ways to think of functions is as the current state of a long battle to conceptualize our world quantitatively. Clagett [9] relates Oresme' s attempts to capture the variational nature of a quality's "intensity" (e.g., temperature) over position and time. Kaput [41] extends Clagett's analysis to trace the evolution of today's ideas of variable and variability in the calculus, concluding that today's static picture of function hides many of the intellectual achievements that gave rise to our current conceptions.

The current standard definition of function highlights correspondence over variation–elements in one set correspond to elements in another so that each element in the first corresponds to exactly one element in the second. Since the 1930s this ordered-pair notion of function has been taken as the "official" definition of function, largely because it solved many problems introduced historically by people like Fourier who wished to define functions by a limit process [31, 44]. The ordered-pair definition has received strong criticism on pedagogical grounds [8, 28, 45, 90] – that it can be meaningful only to people who recognize the problems it solves, but not to a student who is new to the idea of function. On the other hand, we can point to many natural occurrences of correspondences that cannot be expressed analytically or imagined as the product of covariation but which we still would like to call functions (e.g., person's name to person's social security number in a relational database), so a non-correspondence understanding of function is too restrictive in regard to relationships we would like say are functional relationships.

The tension between thinking of function as covariation and of function as correspondence is natural. They are both part of our intellectual heritage, so they show up in our collective thinking. Poincaré put the matter nicely when he said:

> Perhaps you think I use too many comparisons; yet pardon still another. You have doubtless seen those delicate assemblages of siliceous needles which form the skeleton of certain sponges. When the organic matter has disappeared, there remains only a frail and elegant skeleton of certain sponges. True, nothing is there except silica, but what is interesting is the form this silica has taken, and we could not understand it if we did not know the living sponge which has given it precisely this form. Thus

it is that the old intuitive notions of our fathers, even when we have abandoned them, still imprint their form upon the logical constructions we have put in their place. (Poincaré, 1913, p. 219, quoted in [**65**] p. 16).

Function as covariation is one of those " old intuitive notions of our fathers" of which Poincaré spoke. It is natural that vestiges of it show up in our mathematical culture. But we still face the question of how to reflect our heritage within a curriculum in a way that is coherent in regard to a conceptual development of the subject and at the same time respects current mathematical conventions. One way is to reflect the historical development within the curriculum–emphasize function as covariation in K-14, and then introduce function as correspondence as the need arises (e.g., differential equations; pointwise and uniform convergence of function sequences). This would also respect current thinking about the development of function conceptions through the levels of action, process, and object.

I wish to mention quickly that in today's K-14 mathematics curriculum there is no emphasis on function as covariation. In fact, there is no emphasis on variation. I examined the most recent editions of two popular K-9 text series and found that the closest they come to examining variation is to have students construct tables of data, and even then there is a profound confusion between the ideas of random variable and variable magnitude.[3] This is in stark contrast to the Japanese elementary curriculum [**37**] which repeatedly provokes students to conceptualize literal notations as representing a continuum of states in dynamic situations.

Finally, I am surprised that so little has been investigated in regard to students' concepts of variable magnitude–the focus instead being on variable as literal representation of number [**1, 79, 87**]. It seems, to me anyway, that a progressively more abstract notion of covariation rests upon a progressively more abstract image of variable magnitude.

Understanding Phenomena and Representing Phenomena

Zorn, in a report of a NSF-sponsored conference on the state of calculus reform, said that common complaints were represented by one physicist's remark that, whether having had calculus or not, " his students were 'as innocent as newborn babes' about acceleration and velocity" ([**91**] p. 1). Many mathematicians respond to this complaint by saying that we teach mathematics, not physics. The larger issue, I believe, is to what extent we should expect our students to understand the "stuff" that mathematics is about in its applications.

---

[3]The confusion is between, for example, my height as it varies over a sample of 1000 moments in time and the heights of a sample of 1000 people. The 1000 measurements of my height are values of a variable magnitude that can be thought to covary with time; the heights of 1000 people at one moment in time can be thought of as values of a random variable or one value of a vector-valued function, but thinking of them either way would not justify speaking of "height as a function of age" as these texts do.

A debate about whether we should teach physics or chemistry in mathematics class will not be productive. A more productive debate will center around the extent to which we orient students toward conceptualizing the situations our mathematics is about at the moments we use it, and, to relate this debate to functions, what role conceptions of function might play in supporting or inhibiting students' conceptualizations of situations.

Introductory calculus is a natural site to begin discussions of situational conceptions in relation to curriculum, pedagogy, and student learning. For most students, it is the first time they meet functions as models of quantitative situations.[4] The research by Monk [50, 51] suggests that students' difficulties with applications run much deeper than their difficulties with the visible mathematics taught in class.

Monk investigated students' conceptualizations of two classical situations having to do with related rates:

(1) A person is walking toward a street lamp; students are asked to relate changes in the length of the person' s shadow to changes in his distance from the lamp [51] .

(2) A ladder is sliding down a wall; students are asked to relate changes in one end's height above the floor to changes in the other end's distance from the wall [50].

Monk provided physical models for students' experimentation and asked questions about the situations that encouraged students to reason with the physical devices. We should take special note of one aspect of students' reasoning in Monk's reports: Their difficulty in developing a coherent conceptualization of a physical model as a system of dependencies among quantities whose values vary–even while holding the devices in their hands and playing with them. For example, in [51] one student, who was one of the more adept at symbolic mathematics, was also quite certain that (in the street lamp situation) the tip of a person's shadow moves at an increasing rate while the shadow's length changes at a constant rate. This suggests to me that whatever students have in mind as they employ symbolic mathematics, it often is not the situation their professors intend to capture with their symbolic mathematics. It also suggests to me that we need to pay much closer curricular and pedagogical attention to students' pre-symbolic actions, such as imagining dynamic situations so that their images adhere consistently to systems of constraints. In [73] I propose that imagining situations as being functionally constituted is also part of seeing generality in geometric diagrams, and that we can actively promote this ability in schoolchildren with carefully crafted curriculum and instruction. Perhaps the same types of activities would be productive for college students.

The importance of attending to students' conceptualizations of situations ap-

---

[4]This is not to say that subject matter in school mathematics could not be cast as involving functions as models. Rather, the instruction received by most students rarely emphasizes even thinking about situations, let alone functions as models of situations.

plies to more than physical phenomena and physical quantities. It applies whenever we use mathematical notation referentially. In informal investigations of
senior mathematics majors' and secondary mathematics teachers' understandings of the problem shown in Figure 4, I have found, in principle, three categories
of conceptualizations.

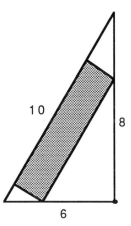

FIGURE 4.   Diagram to accompany the problem, "A rectangle is
situated within a triangle as shown above. Find dimensions of the
rectangle that maximize its area."

The first is that they imagine moving one corner of the rectangle and all other
parts adjust automatically to conform to the problem's constraints (Figure 5a).
These students tend to express functional descriptions of area in relation to
lengths of side 1 and side 2, and to express the lengths of side 1 and side 2 in
relation to the length of some segment. Another category is reflected in Figure
5b. It is that the initial rectangle is made into other rectangles by moving each
corner independently of the others. These students don't even reach a point
where they consider what quantity area might be a function of– if they attempt
to think of area as a function of anything then they are obstructed by the fact
that, in their conception, it is a function of four variables. The third category
(Figure 5c) is that they focus their attention on one side, imagining they are
moving the rectangle much as do students in the first category, but they do not
attend to anything but the side as a whole. They were obstructed from conceiving
area as a function of a something by the fact that, within their current conception
of the situation, they felt there was nothing they could quantify as a measure of
the aspect they imagined themselves manipulating.[5]

A pedagogical implication of these examples is that, for students whose conceptualizations fall into the second two categories illustrated in Figure 5, an

---

[5] I included in the first category those students who thought of moving a side while thinking
of measuring a distance between a vertex on the triangle and a vertex on the rectangle.

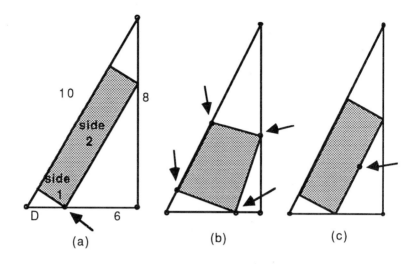

FIGURE 5. Three categories of conceptualizing the underlying situation: (a) Move one vertex, everything moves accordingly, lengths of sides, and hence area, are a function of the distance from vertex D to corner of rectangle. (b) Move each of the rectangle's corners to get another rectangle. (c) Move a side of the rectangle; everything else moves accordingly. Area is somehow a function of "where the side is."

instructor who fails to understand how they are thinking about the situation will probably speak past their difficulties. Any symbolic talk that assumes students have an image like that in the first category will not communicate. These students need a different kind of remediation, a remediation that orients them to construct the situation as one of constrained variation. Only then will they be in a position to understand the task as originally intended, to represent analytically a covariation of magnitudes.

The examples given by Monk, as described previously, and the example I gave above should not be dismissed as somehow pathological. A growing body of evidence suggests that this kind of miscommunication– instructors erroneously assuming students have a principled understanding of an underlying situation– is far from uncommon. Alba Thompson and I have found this to be the case in the teaching of rate in middle school [67, 68] and in the teaching of calculus [75]. Research by White and by Ueno [78, 88, 89] suggests quite strongly that this type of miscommunication also occurs frequently in the teaching of physics.

Operations on Numbers and Operations on Functions

The process conception of function described by Dubinsky and his colleagues [2, 6, 24] emphasizes arithmetic operations as operations on numbers, so that, for example $f(x) = x^2 + 3x$ is the function determined by evaluating the sum of a number squared and three times the number. This seems non problematic. We can even consider the family of functions $f_a(x) = x^2 + ax$. A common exercise in secondary school algebra is to show that the graphs of any function in this family is a translation of the graph of another (Figure 6).

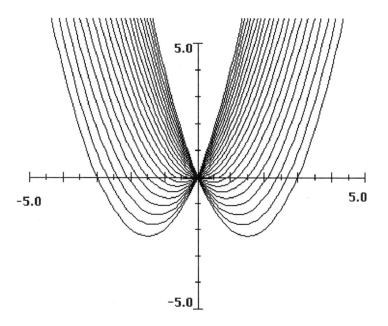

FIGURE 6.   The family of functions $f_a(x) = x^2 + ax$ as a ranges from -3 to 3 in increments of 0.3. Each graph in the family is a translation of any other graph in the family.

An article by Dugdale [27] inspired me to ask students in one course for senior mathematics majors to examine the influence that changes in the linear coefficient has on the behavior of functions in the family $g_a(x) = x^2 + ax$, $(a, x \in \mathbb{R}, n \in \mathbb{N}$. They quickly discovered that thinking of one graph as being the translation of another gave little insight into the general effect that changing the value of a has for $n > 2$ (Figure 7).

A little reflection makes their difficulty clear. Thinking of the graph of one quadratic as being a translation of another draws only on pointwise correspondence of points in the Cartesian plane. It is somewhat coincidental that, in the case of two quadratics f and g, there exist numbers a and b that will relate two subsets of the plane defined by $\{(x, y)|\ y = f(x)\}$ and $\{(u, v)|\ v = g(u)\}$ so that $f(x) + b = g(x + a)$. This relationship does not generalize to polynomials of degree greater than 2. It was only after these students came to think of the expression $xn + ax$ as a sum of functions instead of as a sum of numbers that they gained insight into the effect of varying the linear coefficient in a way that

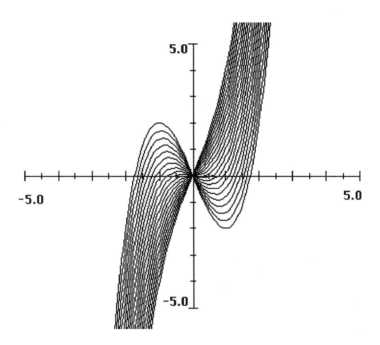

FIGURE 7. The family of functions $f_a(x) = x^3 + ax$ as a ranges from
-3 to 3 in increments of 0.3. The graphs are *not* translations of each
other.

generalized across n. Figure 8 shows their generalized image of the effect of
varying a in the case of. By varying the linear coefficient we change the slope
of the line upon which segments of length x^2 are placed, but we continue to add
the same values of x^2. Seeing the effect this way makes it clear why the family
appears as it does in Figure 7 — the values of x^3 remain unchanged, but are
being "attached" to a line of varying slope.

I suspect that an orientation toward viewing arithmetic operations as operations on numbers supports students natural inclination to view graphs as
pictorial objects *sans* points. I often hear even mathematics majors speak of
a graph as "stretching" or "getting skinnier" or "being squished" without any
thought being given to an underlying dynamics of functional relationship. Goldenberg [30] notes that a similar orientation toward casting change of scale and
change of axes as operations on functions tends to direct students away from
thinking of functional relationship and toward thinking of graphs as objects in
and of themselves.[6]

The notion of derivative and integral as operators (e.g., on products, sums,

---

[6]One reviewer interpreted this comment as being denigrative of graphs as mathematical
objects. This is not the case. Rather, I am speaking about students thinking of graphs as
*non*mathematical objects—as if they were a piece of string or a rubber band. I see no benefit
in students holding such conceptions of graphs.

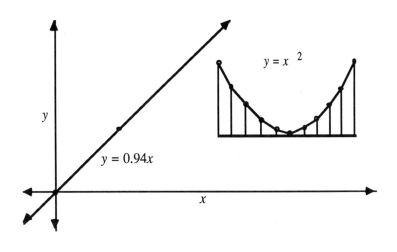

(a)

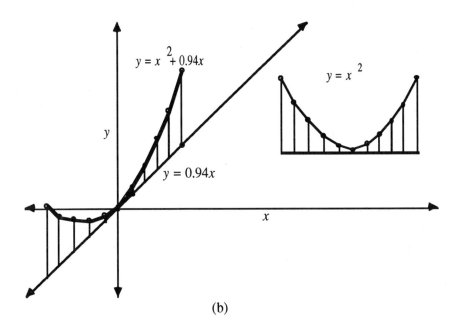

(b)

FIGURE 8.   (a) The functions $f(x) = 0.94x$ and $g(x) = x^2$. (b) The function $h(x) = x^2 + 0.94x$ as the sum of $f(x)$ and $g(x)$. Varying the linear coefficient varies the slope of the graph of f, but the increments due to g remain constant—they just get moved up or down as the linear coefficient varies.

quotients, compositions of functions) is based on seeing expressions as being comprised of operations on functions. A curricular and instructional emphasis in algebra and precalculus on having students develop images of arithmetic operations in analytically-defined functions as operations on functions would seem to prepare them for a deeper understanding of this aspect of the calculus. At the same time, a conception of operations in expressions as operating on numbers and not on functions would seem to be an obstacle to understanding the derivative and integral as linear operators. These are empirically testable hypotheses; I would welcome research on them. Another investigation suggested by this line of argument would be to assess the conceptual requirements for understanding expressions as being comprised of operations on functions. I suspect they would entail, at least, an image of function as process completed over a domain. But, again, this is an empirically testable hypothesis–one which would be ideally suited for a research/development project, since to investigate this question one would need to develop curriculum and pedagogy aimed at having students learn to think of functions as something to be operated on.

Emergent Issues

A number of issues emerge from the literature on function concepts that have not been directly researched, but nevertheless seem important. These are:

• Are specific kinds of intellectual operations required to conceptualize different kinds of functions?

• To what extent are students difficulties a product of instructional obstacles?

• To what extent are students constrained by our misunderstanding their practical realities? ·

Development of specific functions. Recent research in students' understanding of multiplicative structures [**16, 17, 74, 81, 82**] suggests that students who develop strong concepts of function begin doing so by building images of quantitative covariation. But at the same time, it is becoming evident that "quantitative covariation" is not a unitary construct. My own research [**74, 76**] suggests that concepts of linear function emerge from deep understandings of rate. Earlier in this paper I argued that understanding polynomial functions entails special conceptualizations. Dreyfus research [**22**] highlights the need for students to conceptualize periodic phenomena if they are to develop more than a superficial understanding of trigonometric functions. The work by Confreys research group [**17, 62**] suggests that students must first comprehend recursive processes in order to conceptualize exponential functions.

The case of exponential functions is especially instructive. I have shared the graph shown in Figure 9 with a number of secondary mathematics teachers and university professors, asking "during what period of time was inflation the greatest?" Responses, vary, but they almost uniformly include the period from 1978 to 1980. When asked to pick the period with the next highest inflation

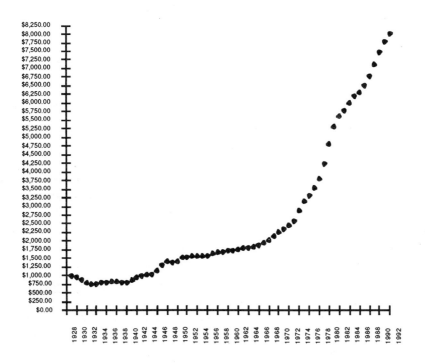

FIGURE 9. Thee price of an item marked in 1929 at $1,000 during the years 1930-1992, adjusted for inflation.

rate, responses vary considerably more–but they have yet to include the period from 1945 to 1948, the period which includes the actual highest inflation rate.

After having been told that 1945-1948 contains the actual highest inflation rate, many people say something like, "Oh, of course–you have to compare one year's price as a percent of the previous year's price." But even more (so far, none have been university professors) could not reconcile my comment with their reading of the graph. It seems that they were looking at inflation as increase in price per year (i.e., rate of increase with respect to time), which translates into slope, instead of as percentage change, which is a recurrence relationship. Confrey and her group argue that the notion of recurrence is common to conceptualizations of situations that entail exponential growth. They also note that this style of thinking is absent in the school mathematics curriculum. I suspect that many calculus instructors routinely assume that it is non-problematic for students to envision exponential growth when it seems few are inclined to do so without a great deal of orientation on the instructor's part.

Our tacit expectations: Cognitive and instructional obstacles.

Herscovics [34] explicated the notion of cognitive obstacle as it relates to learning mathematics. A cognitive obstacle is a way of knowing something that gets in the way of understanding something else. For example, thinking of a graphics construction as producing only what is immediately envisioned can be an obstacle in regard to conceptualizing the construction of a fractal [**70**]. An instructional obstacle is instruction that promotes new cognitive obstacles or supports or is neutral in regard to students' existing cognitive obstacles. Sierpinska [**61**] discusses a number of instructional obstacles to the understanding of functions. I will attempt to illustrate how we often contribute unthinkingly to our students' difficulties.

Many calculus texts begin their section on implicit differentiation with a discussion of how an equation in two variables somehow hides a function-one variable being a function of the other. *Behind this approach lurk three conceptions of function, and students are not alerted that there is something subtle going on.* The first conception of function is of a multi-variable equation$f(x, y) = c$. That is, we are looking at the pre-image of a level curve. The second conception is that of function as rule. It is suggested that from the equation we can, or at least would like to, rewrite the equation to get a rule for obtaining values of one variable from values of the other. The third conception is that of function as correspondence between sets. Any subset of the pre-image that determines a univocal mapping from one variable to the other is a function.

My point in relating this common opening to implicit differentiation is twofold: First, it is often evident that what the author really has in mind is getting to the rule for implicit differentiation with no real expectation that students understand the notion of implicit function. Second, if the intent is for students to understand the setting under which implicit differentiation happens, then they should be alerted to how complex the setting is. "Warning, Go slow! We're going to look at a pretty sophisticated set of ideas here."

I entitled this section "Our tacit expectations." Do we expect students to really understand what we teach? If no, then we need to say so. If yes, then we need to expect understanding, communicate that expectation, and provide curricular and pedagogical support for students to meet our expectations.

Students' practical realities. We must keep in mind that our college students have spent 12 years in school learning that mathematics is a ritualistic behavior, and that often their expectation of us is to "show them how to do it." Their mathematical experiences did not include learning how to use notation thoughtfully and reflectively; notation is something to be seen, not to be interpreted. But, to change students' orientation requires a "renegotiation of the didactic contract" [**3, 49**]. Students need to know that we know where they are coming from, and that ritualistic performance is not satisfactory. At the same time, we must assume the responsibility for shaping our instruction so that it *can* be understood conceptually, and to do that we must attend constantly to matters of imagery and understanding.

Multiple Representations

While the importance of students' understanding expressions, tables, and graphs has been common knowledge for at least a century, it is only since the early 1980's that they have been seen cognitively and pedagogically as alternative windows on a central idea [**38, 39, 42**]. Even though a semblance of multiple representations can be seen in Diene's original idea of multiple embodiments of mathematical concepts [**20, 47**], the notion of multiple representations has today become a powerful motor of curricular research and development largely because of access to increasingly powerful computers and graphing calculators. I refer you to [**57**] and [**46**] for state-of-the-art reviews of research on graphs, tables, and expressions as they relate to the matter of multiple representations of function. I will instead give a cautionary note regarding an important missing element in this line of research and development.

I believe that the idea of multiple representations, as currently construed, has not been carefully thought out, and the primary construct needing explication is the very idea of representation.[7] Tables, graphs, and expressions might be multiple representations of functions to us, but I have seen no evidence that they are multiple representations of anything to students. In fact, I am now unconvinced that they are multiple representations even to us, but instead may be areas of representational activity among which, as Moschkovich, Schoenfeld, and Arcavi [**52**] have said, we have built rich and varied connections. It could well be a fiction that there is any interior to our network of connections, that our sense of "common referent" among tables, expressions, and graphs is just an expression of our sense, developed over many experiences, that we can move from one type of representational activity to another, keeping the current situation somehow intact. Put another way, the core concept of "function" is not *represented* by any of what are commonly called the multiple representations of function, but instead our making connections among representational activities produces a subjective sense of invariance.

I do not make these statements idly, as I was one to jump on the multiple-representations bandwagon early on [**71, 72**], and I am now saying that I was mistaken. I agree with Kaput [**40**] that it may be wrongheaded to focus on graphs, expressions, or tables as representations of function. We should instead focus on them as representations of something that, from the students' perspective, is *representable,* such as aspects of a specific situation. The key issue then becomes twofold: (1) To find situations that are sufficiently propitious for engendering multitudes of representational activity and (2) To orient students toward drawing connections among their representational activities in regard to the situation that engendered them. The situation being represented must be paramount in students' awareness, for if they do not see something remaining the same as they move among tables, graphs, and expressions, then it increases the probability that they will see each as a "topic" to be learned in isolation of the others. Dugdale [**26**] provides an excellent example of a productive and

---

[7]This is entirely parallel to the situation in information processing psychology—no one has bothered to question what is meant by "information"[**10, 11**].

powerful coordination of situation and representation.

Reflections

Much of the literature on students' concepts of function highlights what they do not know about functions and why that might be the case. One lesson we can learn is that students can use abstract function concepts only if they build connected abstractions from "functional" reasoning–reasoning about constrained covariation, reasoning about representations of quantitative and numerical relationships, and reasoning about properties of relationships. Curricula that are constructed logically, but which do not attend to transitional conceptualizations, can put students at risk of having to cope with demands of performance by turning our representations into their objects of learning [48, 59, 60].

Another lesson to draw from the foregoing is that our success in vitalizing the undergraduate curriculum is highly dependent upon a great deal more conceptual development happening in schools. If students come to us with impoverished concepts of function, then there is not a lot we can do except accommodate to the constraints we find.[8] We must support, collectively and individually, the efforts of NCTM [14, 15] to reform school mathematics. We cannot succeed unless they succeed.

Finally, we need to broaden our notion of appropriate curriculum. Logical developments of content look especially good to people who already know it, but they are "logical" precisely because they express the logic we have constructed in our understandings of the subject. I propose that we orient ourselves toward developing *conceptual* curricula–curricula that are mathematically sound, but nevertheless are constructed from the start with an eye to building students' understandings, and are constructed to assess skill as an expression of understanding. I take this as our major challenge over the coming decade.

REFERENCES

1. A. Arcavi and A. Schoenfeld, *On the meaning of variable*, Mathematics Teacher **81**(6) (1987), 420–427.
2. T. Ayers, et al., *Computer experiences in learning composition of function*, Journal for Research in Mathematics Education **19**(3) (1988), 246–259.
3. N. Balacheff, *Towards a problmatique for research on mathematics teaching*, Journal for Research in Mathematics Education **21**(4) (1990), 258-272.
4. J. Bamberger and D. A. Schn, *Learning as reflective conversation with materials*, Research and reflexivity, F. Steier, Editor, Sage, London, 1991.
5. H. Bauersfeld, *Hidden dimensions in the so-called reality of a mathematics classroom*, Educational Studies in Mathematics **11**(1) (1980), 23–42.
6. D. Breidenbach, et al., *Development of the process conception of function*, Educational Studies in Mathematics (in press).

---

[8]This is not to say that we cannot do anything. It means that whatever we try must take into account the low level of experience in grounded, functional reasoning which our students will commonly bring to the classroom.

7. L. E. J. Brouwer, *Consciousness, philosophy, and mathematics*, Proceedings of the Tenth International Congress of Philosophy (1949), North-Holland Publishing Co., Amsterdam, The Netherlands.

8. R. C. Buck, *Functions, in Mathematics education:*, Sixty-ninth yearbook of the National Society for the Study of Education, E. G. Begle, Editor (1970), University of Chicago Press, Chicago, IL.

9. M. Clagett, *Nicole Oresme and the medieval geometry of qualities and motions.* (1968), University of Wisconsin Press, Madison, WI.

10. P. Cobb, *Information-processing psychology and mathematics education: A constructivist perspective*, Journal of Mathematical Behavior **6** (1987), 3–40.

11. P. Cobb, *A constructivist perspective on information-processing theories of mathematical activity*, International Journal of Educational Research **14** (1990), 67–92.

12. P. Cobb, et al., *Characteristics of classroom mathematics traditions: An interactional analysis*, American Educational Research Journal **29(3)** (1992), 573–604.

13. K. F. Collis, *Concrete operational and formal operational thinking in mathematics*, The Australian Mathematics Teacher **25(3)** (1969), 77–84.

14. Commission on Standards for School Mathematics, *Curriculum and evaluation standards for school mathematics*, National Council of Teachers of Mathematics (1989), Reston, VA.

15. Commission on Standards for School Mathematics, Professional standards for teaching mathematics, National Council of Teachers of Mathematics (1991), Reston, VA.

16. J. Confrey, *Splitting, similarity, and rate of change: A new approach to multiplication and exponential functions*, The development of multiplicative reasoning in the learning of mathematics, G. Harel and J. Confrey, Editor, SUNY Press, Albany, NY, in press.

17. J. Confrey and E. Smith, *Exponential functions, rates of change, and the multiplicative unit.*, Paper presented at the Annual Meeting of the American Educational Research Association, San Francisco, April 1992.

18. R. G. Cooper, *The role of mathematical transformations and practice in mathematical development*, Epistemological foundations of mathematical experience, L. P. Steffe, Editor, Springer-Verlag, New York, 1991, pp. 102–123.

19. R. B. Davis, E. Jockusch, and C. McKnight, *Cognitive processes involved in learning algebra*, Journal of Children's Mathematical Behavior **2(1)** (1978), 10–32.

20. Z. Dienes, *Building up mathematics. 4th ed.*, Hutchinson Educational, London, 1960.

21. R. G. Douglas, ed., *Toward a lean and lively calculus*, MAA Notes **6.** (1986), Mathematical Association of America, Washington, D. C..

22. T. Dreyfus and T. Eisenberg, *On teaching periodicity*, International Journal of Mathematical Education in Science **11** (1980), 507-509.

23. E. Dubinsky, *Reflective abstraction in advanced mathematical thinking*, Advanced mathematical thinking, D. Tall, Editor, Kluwer, Dordrecht, The Netherlands, 1991, pp. 95–123.

24. E. Dubinsky and G. Harel, *The nature of the process conception of function*, The concept of function: Aspects of epistemology and pedagogy, G. Harel and E. Dubinsky, Editor, Mathematical Association of America, Washington, DC, 1992, pp. 85–106.

25. E. Dubinsky and K. Schwingendorf, *Calculus, concepts, and computers (preliminary version)*, West Publishing, St. Paul, MN, 1992.

26. S. Dugdale,, *Functions and graphs: Perspectives on student thinking*, Integrating research on the graphical representation of functions, T. A. Romberg, E. Fennema, and T. P. Carpenter, Editor, Erlbaum, Hillsdale, NJ, 1993, pp. 101–130.

27. S. Dugdale, L. J. Wagner, and D. Kibbey, *Visualizing polynomial functions: New insights from an old method in a new medium*, Journal of Computers in Mathematics and Science Teaching **11(2)** (1992), 123–142.

28. T. Eisenberg, *Functions and associated learning difficulties*, Advanced mathematical thinking, D. Tall, Editor,, Kluwer, Dordrecht, The Netherlands, 1991, pp. 140–152.

29. E. P. Goldenberg, *Mathematics, metaphors, and human factors: Mathematical, technical, and pedagogical challenges in the educational use of graphical representation of functions*, Journal of Mathematical Behavior **7** (1988), 135–173.

30. E. P. Goldenberg, P. Lewis, and J. O'Keefe, *Dynamic representation and the development of a process understanding of function*, The concept of function: Aspects of epistemology

and pedagogy, G. Harel and E. Dubinsky, Editor, Mathematical Association of America, Washington, D. C., 1992, pp. 235–260.

31. E. Gonzlez-Velasco, *Connections in mathematical analysis: The case of Fourier Series*, The American Mathematical Monthly **99(5)** (1992), 427–441.

32. J. G. Greeno, *Number sense as situated knowing in a conceptual domain*, Journal for Research in Mathematics Education **22** (1991), 170–218.

33. G. Harel and E. Dubinsky, ed., *The concept of function: Aspects of epistemology and pedagogy*, MAA Notes **25** (1991), Mathematical Association of America, Washington, DC.

34. N. Herscovics, *Cognitive obstacles encountered in the learning of algebra*, Research issues in the learning and teaching of algebra, S. Wagner and C. Kieran, Editor, Erlbaum, Hillsdale, NJ, 1989, pp. 60-86.

35. N. Herscovics and C. Kieran, *Constructing meaning for the concept of equation*, Mathematics Teacher **73** (1980), 572–580.

36. A. Heyting, *Intuitionism: An introduction. Studies in logic and the foundations of mathematics*, North-Holland Publishing Co., Amsterdam, The Netherlands, 1956.

37. Japanese Ministry of Education, Japanese Grades 7, 8, and 9 Mathematics, *UCSMP Textbook Translations*, ed. K. Kodaira., [Hiromi Nagata Trans.] University of Chicago School Mathematics Project, Chicago, IL, 1992.

38. J. J. Kaput, *Representation and problem solving: Methodological issues related to modeling*, Teaching and learning mathematical problem solving: Multiple research perspectives, E. A. Silver, Editor, Erlbaum, Hillsdale, NJ, 1985, pp. 381–388.

39. J. J. Kaput, *Toward a theory of symbol use in mathematics*, Problems of representation in mathematics learning and problem solving, C. Janvier, Editor, Erlbaum, Hillsdale, NJ, 1987.

40. J. J. Kaput, *The urgent need for proleptic research in the graphical representation of quantitative relationships*, Integrating research in the graphical representation of functions, T. Carpenter, E. Fennema, and T. Romberg, Editor, Erlbaum, Hillsdale, NJ, 1993, pp. 279–311.

41. J. J. Kaput, *Democratizing access to calculus: New Routes to old roots*, Mathematics and Cognitive Science, A. H. Schoenfeld, Editor, Mathematical Association of America, Washington, D. C., in press.

42. J. J. Kaput, et al., *The role of representations in reasoning with intensive quantities: Preliminary analyses. Technical Report 869*, Cambridge, MA, Harvard University, Educational Technology Center (1986).

43. C. Kieran, *Functions, graphing, and technology: Integrating research on learning and instruction*, Integrating research in the graphical representation of functions, T. Carpenter, E. Fennema, and T. Romberg, Editor, Erlbaum, Hillsdale, NJ, 1993, pp. 189–237.

44. I. Kleiner, *Evolution of the function concept: A brief survey*, College Mathematics Journal **20** (1989), 282–300.

45. M. Kline, *Logic versus pedagogy*, The American Mathematical Monthly **77(3)** (1970), 264-282.

46. G. Leinhardt, O. Zaslavsky, and M. K. Stein, *Functions, graphs, and graphing: Tasks, learning, and teaching*, Review of Educational Research **60(1)** (1990), 1–64.

47. R. Lesh, T. Post, and M. Behr, *Dienes revisited: Multiple embodiments in computer environments*, Developments in school mathematics education around the world, I. Wirszup and R. Streit, Editor, National Council of Teachers of Mathematics, Reston, VA, 1987, pp. 1–25.

48. L. Linchevski and A. Sfard, *Rules without reasons as processes without objects*, Proceedings of the Fifteenth Annual Conference of the International Group for the Psychology of Mathematics Education (1990), Assisi, Italy.

49. J.-J. Lo, G. H. Wheatley, and A. C. Smith, *Negotiation of social norms in mathematics learning*, Proceedings of the Annual Meeting of the International Group for the Psychology of Mathematics Education–North America (1991), Virginia Tech, Blacksburg, VA.

50. G. S. Monk, *Students' understanding of a function given by a physical model*, The concept of function: Aspects of epistemology and pedagogy, G. Harel and E. Dubinsky, Editor, Mathematical Association of America, Washington, D. C., 1992, pp. 175–194.

51. G. S. Monk, *A study of calculus students' constructions of functional situations: The case of the shadow problem.*, Paper presented at the Annual Meeting of the American Educational Research Association, San Francisco, CA, April, 1992.

52. J. Moschkovich, A. H. Schoenfeld, and A. Arcavi, *Aspects of understanding: On multiple perspectives and representations of linear relations and connections among them*, Integrating research on the graphical representation of functions, T. A. Romberg, E. Fennema, and T. P. Carpenter, Editor, Erlbaum, Hillsdale, NJ, 1993, pp. 69–100.

53. J. Piaget, *Genetic epistemology*, W. W. Norton, New York, 1971.

54. J. Piaget, *The psychogenesis of knowledge and its epistemological significance*, Language and learning: The debate between Jean Piaget and Noam Chomsky, M. Piatelli-Palmarini, Editor,, Harvard University Press, Cambridge, MA, 1980, pp. 23–34.

55. J. Piaget and R. Garcia, *Psychogenesis and the history of science. [H. Feider Trans.]* (1989), Columbia University Press, New York.

56. J. Richards, *Mathematical discussions*, Radical constructivism in mathematics education, E. von Glasersfeld, Editor,, Kluwer, The Netherlands, 1991, pp. 13–51.

57. T. A. Romberg, E. Fennema, and T. P. Carpenter, ed., *Integrating research on the graphical representation of functions*, Erlbaum, Hillsdale, NJ, 1993.

58. A. Sfard, *The development of algebra: Confronting historical and psychological perspectives*, Paper presented at the Seventh International Congress on Mathematical Education, Quebec City, Canada, August 1992.

59. A. Sfard and L. Linchevski, *The gains and the pitfalls of reification: The case of algebra*, Educational Studies in Mathematics (in press).

60. A. Sfard and L. Linchevsky, *Equations and inequalities: Processes without objects?*, Proceedings of the Sixteenth Annual Conference of the International Group for the Psychology of Mathematics Education (1992), University of New Hampshire, Durham, NH.

61. A. Sierpinska, *On understanding the notion of function*, The concept of function: Aspects of epistemology and pedagogy, G. Harel and E. Dubinsky, Editor, Mathematical Association of America,, Washington, D. C., 1992, pp. 25–58.

62. E. Smith and J. Confrey, *Multiplicative structures and the development of logarithms: What was lost by the invention of function*, The development of multiplicative reasoning in the learning of mathematics, G. Harel and J. Confrey, Editor, SUNY Press, Albany, NY, in press.

63. L. A. Steen, ed., *Calculus for a new century: A pump, not a filter.*, MAA Notes **8** (1987), Mathematical Association of American, Washington, DC.

64. D. Tall, ed, *Advanced mathematical thinking*, Kluwer, Dordrecht, The Netherlands, 1991.

65. D. Tall, *The psychology of advanced mathematical thinking*, Advanced mathematical thinking, D. Tall, Editor, Kluwer, Dordrecht, The Netherlands, 1991, pp. 3–21.

66. D. Tall and S. Vinner, *Concept images and concept definitions in mathematics with particular reference to limits and continuity*, Educational Studies in Mathematics **12** (1981), 151–169.

67. A. G. Thompson and P. W. Thompson, *Talking about rates conceptually, Part I: A teacher's struggle*, Journal for Research in Mathematics Education (in press).

68. _____, *Talking about rates conceptually, Part II: Pedagogical content knowledge*, Journal for Research in Mathematics Education (in press).

69. P. W. Thompson, *Experience, problem solving, and learning mathematics: Considerations in developing mathematics curricula*, Teaching and learning mathematical problem solving: Multiple research perspectives, E. Silver, Editor, Erlbaum, Hillsdale, NJ, 1985, pp. 189–243.

70. P. W. Thompson, *Understanding recursion: Process $\equiv$ Object*, Proceedings of the 7th Annual Meeting of the North American Group for the Psychology of Mathematics Education (1985), Ohio State University, Columbus, OH.

71. P. W. Thompson, *Mathematical microworlds and intelligent computer-assisted instruction*, Artificial Intelligence and Education, G. Kearsley, Editor, Addison-Wesley, New York, 1987, pp. 83–109..

72. P. W. Thompson, *Artificial intelligence, advanced technology, and learning and teaching algebra*, Research issues in the learning and teaching of algebra, C. Kieran and S. Wagner,

Editor, Erlbaum, Hillsdale, NJ, 1989, pp. 135-161.

73. P. W. Thompson, *Constructivism, cybernetics, and information processing: Implications for research on mathematical learning*, Constructivism in education, J. Gale and L. P. Steffe, Editor, Erlbaum, Hillsdale, NJ.

74. P. W. Thompson, *The development of the concept of speed and its relationship to concepts of rate*, The development of multiplicative reasoning in the learning of mathematics, G. Harel and J. Confrey, Editor, SUNY Press, Albany, NY, in press.

75. P. W. Thompson, *Images of rate and operational understanding of the Fundamental Theorem of Calculus*, Educational Studies in Mathematics (in press).

76. P. W. Thompson and A. G. Thompson, *Images of rate*, Paper presented at the Annual Meeting of the American Educational Research Association, San Francisco, CA, April 1992.

77. U. Treisman, *Studying students studying calculus: A look at the lives of minority mathematics students in college*, The College Mathematics Journal **23(5)** (1992), 362–372.

78. N. Ueno, N. Arimoto, and A. Yoshioka, *Learning physics by expanding the metacontext of phenomena*, Paper presented at the Annual Meeting of the American Educational Research Association,, San Francisco, April 1992.

79. Z. Usiskin, *Conceptions of school algebra and uses of variables*, Ideas of algebra: K-12, A. Coxford, Editor,, NCTM, Reston, VA, 1988, pp. 8–19.

80. D. van Dalen, ed., *Brouwer's Cambridge lectures on intuitionism*, Cambridge University Press, Cambridge, UK, 1981.

81. G. Vergnaud, *Multiplicative structures*, Acquisition of mathematics concepts and processes, R. Lesh and M. Landau, Editor, Academic Press, New York, 1983, pp. 127–174.

82. G. Vergnaud, *Multiplicative structures, in Number concepts and operations in the middle grades, J. Hiebert and M. Behr, Editor*, National Council of Teachers of Mathematics, Reston, VA,, 1988, pp. 141–161.

83. S. Vinner, *The role of definitions in the teaching and learning of mathematics*, Advanced mathematical thinking, D. Tall, Editor, Kluwer, Dordrecht, The Netherlands, 1991, pp. 65–81.

84. S. Vinner, *The function concept as a prototype for problems in mathematics learning*, The concept of function: Aspects of epistemology and pedagogy, G. Harel and E. Dubinsky, Editor, Mathematical Association of America, Washington, D. C., 1992, pp. 195–214.

85. S. Vinner and T. Dreyfus, *Images and definitions for the concept of function*, Journal for Research in Mathematics Education **20** (1989), 356–366.

86. E. von Glasersfeld, *Abstraction, re-presentation, and reflection: An interpretation of experience and Piaget's approach*, Epistemological foundations of mathematical experience, L. P. Steffe, Editor, Springer-Verlag, New York, 1991, pp. 45–65.

87. S. Wagner, *Conservation of equation and function under transformations of variable*, Journal for Research in Mathematics Education **12** (1981), 107–118.

88. B. White, *Intermediate causal models: The missing link for successful science education?,*, Advances in instructional psychology, R. Glaser, Editor,, Erlbaum, Hillsdale, NJ, in press.

89. B. White, *ThinkerTools: Causal models, conceptual change, and science education*, Cognition and Instruction (in press).

90. R. Wilder, *The role of axiomatics in mathematics*, American Mathematical Monthly **74** (1967), 115–127.

91. P. Zorn, *Content and discontent in San Antonio*, UME Trends **3(1)** (1991).

DEPARTMENT OF MATHEMATICAL SCIENCES AND CENTER FOR RESEARCH IN MATHEMATICS AND SCIENCE EDUCATION, SAN DIEGO STATE UNIVERSITY

CBMS Issues in Mathematics Education
Volume 4, 1994

On Understanding How Students Learn
to Visualize Function Transformations

THEODORE EISENBERG & TOMMY DREYFUS

If it does nothing else, undergraduate mathematics should help students to develop function sense–a familiarity with examining relations among variables (Everybody Counts, [13]).

1. INTRODUCTION: THE LARGER CONTEXT

One of the major goals of the school curriculum is to get students to think mathematically. Although each of us probably has a visceral feeling of what it means to think mathematically, we would most likely find it an impossible task to define it in an inclusive and exclusive way. In all likelihood we would be forced to define this educational construct by way of examples; these problems elicit mathematical thinking, those do not. Mathematical thinking is a rubric to characterize a sort of flexibility of thought, a flexibility which we hope develops in students as a result of the curriculum (Mason, Burton and Stacey, [19]). Having a sense for number and having a sense for functions are among the most important facets of mathematical thinking.

1.1 Number sense. A student has a *sense for numbers* if he or she can:

1. Immediately find the largest number among a, b, c, d, given that

$$a - 2 = b + 3 = c - 4 = d - 1.$$

2. Explain why $(354)(271) \neq 905934$.
3. Determine the number of zeros at the end of 100!
4. Find the units digit in the expansion of 347^{286}.
5. Estimate the average speed one has jogged if 12.6 km were covered in 37 minutes.

Although the above problems can also be done by calculations and algebraic techniques, such an approach misses the essence of number sense. Having a sense for numbers encourages a certain flexibility in thinking about numbers, and the linking together of many number concepts which are generally taught in isolation in the school curriculum (Arithmetic Teacher, [1]; Greeno, [14]).

1.2 Function sense. Just as a sense for numbers consists of insights into number properties, operations with numbers and their links, developing function sense implies acquiring analogous insights with respect to functions as they are usually encountered in the high school and undergraduate curriculum. Facets of function sense include dependence (the different roles of the dependent and the independent variable in a function), variation (e.g., increase), co-variation (of two variables), variation of variation (concave means growing at a growing rate) as well as a good grasp of the effects of various operations on functions such as their addition, composition, shift or integration. A *sense for functions* manifests itself, for example, in one's ability to:

1. Find the number of real zeroes in $ax^2 + bx + c = 0$ $(a, b, c$ real), if it is known that $a + b + c > 0$ and $36a + 6b + c < 0$.

2. Find all x where $f(x) < g(x)$ if $f(x) = \sqrt{x+5}$ and $g(x) = x - 1$.

3. Realize that there can be no solutions to $x^2 - \sin(x) + 1 = 0$, and have no trouble counting the number of solutions to $\sin(x) = x$.

4. Describe $ga(x) = ax(2-x)$ as the family of all quadratics/parabolas whose zeros are 0 and 2.

5. Find the minimum of $y = a^2 + b$, $(a > 0)$ without taking the derivative.

6. If f is increasing, and g is decreasing, determine without differentiating whether or not the function given by $f(x) \pm g(x)$ is increasing or decreasing?

7. Immediately determine that the area trapped between the curves given by $y = x^2$ and $y = x$ is the same as the area trapped between the curves given by $y - 1 = (x-1)^2$ and $y - 1 = x - 1$.

8. Explain why $\int_a^b g(x)\, dx = \int_{a+k}^{b+k} g(x-k)\, dx$.

9. Given the graph of a continuous function f, sketch the graphs of related functions such as those given by $f(kx)$, $f(x) + k$, $\dfrac{1}{f(x)}$, $f^{-1}(x)$, $f(|x|)$ and $f'(x)$.

A central aspect to function sense is the use of more than one representation for the same mathematical situation. In fact, multiple representations are so ubiquitous in functions that they can hardly ever be avoided. The ability to pass from one representation to another when appropriate, the flexibility to use the most appropriate representation in solving a problem, the ability to "see" one representation when working in another is one of the most essential components of function sense (Schwarz, [**25**]). One important representation of real functions of a real variable is the Cartesian graph and, indeed, the key to solving many of the above problems is to think of them visually. Take for example the statement that $\int_a^b g(x)\, dx$ must necessarily equal $\int_{a+k}^{b+k} g(x-k)\, dx$. One can justify this by making an appropriate substitution in the integral. But one can also visualize that the graph of g has been shifted k units (to the right if $k > 0$) and so must the limits of integration. We believe that having the ability to visualize the graph of a function and elementary operations and notions on it is an important aspect of having a sense for functions. It contributes in an essential way to many of the aspects of function sense mentioned above, among them variation, variation of variation, differentiation, shift, stretch and other transformations.

1.3 Visualization. It is well known, on the other hand, that visualization has its drawbacks. For example, Presmeg [23] found that *the "stars" in senior high school classes [are] almost always nonvisualizers* (p. 300). Her conclusion was based on two results: (i) that many visualizers were not aware how to transcend the one-case concreteness of an image or a diagram, and (ii) that the school curriculum reinforces those who prefer to use non visual methods. Presmeg left as an open question whether or not this (non visual) pattern of thinking is adequate at higher levels when creative mathematical thought is required.

A related result, which is also well documented in the literature is that even advanced students are reluctant to think visually. They often gravitate away from visual paths of information processing toward algebraic ones, even when the latter are more complicated (Clements, [4]; Eisenberg and Dreyfus, [12]; Vinner, [30]; Mundy, [20]). We hypothesize that one of the reasons students do not feel comfortable with visualizations is because they have not constructed cognitive frameworks in which to think of them. Although many of the above problems can be thought of in graphical terms, students seem reluctant to do so. They consider graphical interpretations to be external to the concept of function itself. To them, graphical interpretations are not integral parts of functions but rather something peripheral to functions. Many students realize that graphs of many functions exist, but they are considered as extra baggage a function carries. We believe that this way of thinking about functions is very restrictive and limits one's sense for functions.

Therefore in view of Presmeg's open question, our belief that having a sense for functions entails the ability to visualize them, and the fact that most students are reluctant to do so, we designed a teaching experiment (in the sense described by Romberg, [24]). The purpose of this experiment was to help students think of functions in a visual way, and to help us understand the obstacles they must overcome in doing so. Because the ability to establish links between visual and algebraic thinking becomes particularly important for function transformations, we chose this topic for our investigation.

2. THE STUDY

2.1 Goals. A study was designed to investigate the effects of a teaching unit whose aim was to get students to think of function transformations in a visual way. Specifically, from a given graph or algebraic description of a function f students were taught to visualize, discuss and graph the result of the transformations $f(x) \to kf(x)$, $f(x) \to f(kx)$, $f(x) \to f(x-k)$, $f(x) \to f(x)+k$, $f(x) \to f(|x|)$, $f(x) \to |f(x)|$ and various combinations of these, including the transformation of the stock parabola given by $f(x) = x^2$ to the transformed parabola given by $f(x) = a(x-d)^2 + e$.

An intensive effort was made to get students to think visually about such transformations. The aim of the instructional unit was to help students adopt visual ways of reasoning, and to establish the connection between the visual representation of function transformations and their algebraic description. The aim of our investigation was, therefore, to study the effects of the instructional unit on getting students to process analytic descriptions of algebraic statements visually.

We were well aware from the outset of the teaching experiment that we were working with a very special class of functions, those which have algebraic and graphical descriptions as opposed, for example, to recursive functions or functions which are stated as procedures. But the vast majority of functions encountered in the high school and collegiate curriculum are of the type we studied, and we wanted to understand how students dealt with them.

2.2 Population. The student population was drawn from a boys' senior high school in Israel. All students in the sample (ages 17/18) were studying mathematics in the highest stream in the curriculum which culminates with an introduction to differential and integral calculus. The national matriculation examination for this stream typically requires students to graph, with calculus tools, functions such as $f(x) = xe^x$ and to find the equation of the straight line through the origin bisecting the area between the curve $y = 6x - x^2$ and the x-axis.

Thirty-one students took a written pretest which contained two parts: Standard algebra questions and non-standard questions on function transformations. This pretest had the form of a questionnaire which is discussed in Section 2.3.B. Students with top and bottom pretest scores were eliminated from the experiment because it was felt that they either already knew the material to be taught or that we could never get them to understand the material within the time constraints placed on the experiment by the school administration. Because of the limited number of computers available, eight pairs were formed from the remaining pool of students. The 16 students finally selected participated in the experiment voluntarily. On the whole, they were competent on standard algebra questions and they had a good grasp of the link between the graphical and the algebraic representations of functions. Their success rate on the corresponding part of the pretest was 83%; more specifically their success rate on Question 10b (see Figure 1, Section 3.3.B) was 88%. On the other hand, they lacked the ability to perform function transformations as tested in the pretest.

There are many reasons why only 16 students participated in this teaching experiment. We threw out of the experiment those whom we thought could not benefit from it because they either knew too much or too little at the outset; we were limited by the number of computers in the school; we were limited by the time constraints placed on us by the school administration, etc. But the purpose of this teaching experiment was to give us insights into how students internalize the notion of transformations, and for this and the methodology adopted, 16 students seemed to be a perfect size.

2.3 Instruments.

2.3.A Software. The Green Globs computer game was developed by Dugdale and Kibbey [9] with the aim of providing an entertaining and exciting framework for furthering the understanding of functions and their graphs. Specifically, the authors state that the program helps students ... *develop various strategies and increase their particular skill in graphing... [and that they can] ... construct interesting graphs beyond those normally studied in high school algebra.* The program presents a number of small circles (globs) in a coordinate system on the screen and the task is for the student to find functions whose graphs pass through as many globs as possible. The standard range for the variables is $-10 < x, y < 10$. The student has to enter an algebraic representation for the function. Points are accumulated for each glob traversed, one point for the first glob, two for the second, four for the third and 2^{n-1} for the nth. The game is played until all globs are traversed: the student types in the equation of a graph, the computer graphs it and accumulates the score; then it is ready to accept the next equation. Graphs which would essentially cover the entire screen such as that given by $y = 10\sin(20x)$ are excluded by imposing a maximal length on the graph. The Green Globs program has won many awards and citations. Education News Service listed Green Globs as being one of the best of the more than 11,000 educational programs they evaluated (Davis, [5]). To us, it seemed to be the best graphing program commercially available at the time of the experiment.

2.3.B Questionnaire. A questionnaire for assessing standard algebra skills and performance on function transformations was constructed. The questionnaire consisted of two parts. The standard part contained 17 questions involving substitutions, solving equations, graphs of parabolas, etc. The non-standard part contained 34 questions which related to the role parameters play in graphs of functions. For example, students had to determine the effect of subjecting the graph of quadratic and more general functions f to the transformations $f(x) \to f(kx)$, $f(x) \to f(x - k)$, $f(x) \to kf(x)$ and $f(x) \to f(x) + k$. Sample questions are given in Figure 1.

2.3.C Interviews. Interview questions were constructed with mathematical content similar to the non-standard part of the questionnaire. The purpose of this instrument was to help us determine the extent to which the program helped students process information on functions and transformations on them visually. The instrument consisted of three questions, each of which was novel to the students and at a higher level than those in the questionnaire.

The interview questions are listed in Figure 2. This instrument required the students to think aloud through all stages of the interview. This was done to identify their thought processes and to determine whether, in solving the function transformation problems, they were processing the information visually or analytically.

Standard questions

1(b). Give an example of a function f which satisfies $f(1) = 0$, $f(2) = 1$, $f(3) = 4$.

2(b). Simplify $(4x^2 - 9y^2) - (x - 3y)(x + 3y)$.

9(a). Solve $\dfrac{(x^3 - x^2)}{x} = 0$.

10(b). Next to each function write the corresponding graph (I, II, III, or IV)

____ $x^2 - 2x + 1$ ____ $1 - x^2$ ____ $x^2 - 2x$ ____ $x^2 + 1$

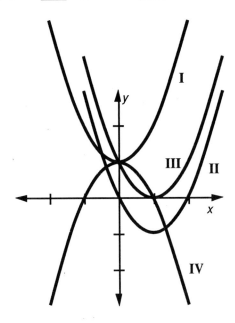

Non-standard questions

3(h). Does the graph below describe a function whose formula is of the form $y = ax + b$ with $a > 1$ and $b > 1$?

4(b). Draw at least three graphs which correspond to a formula of the form $y = ax^2 - a$, $a > 0$.

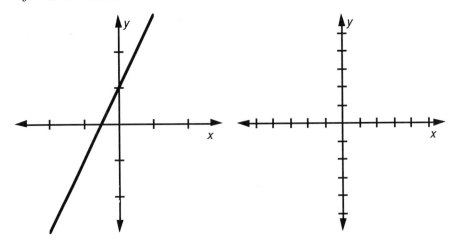

5(b). In the figure on the left you are given the graph of the function $y = f(x)$.

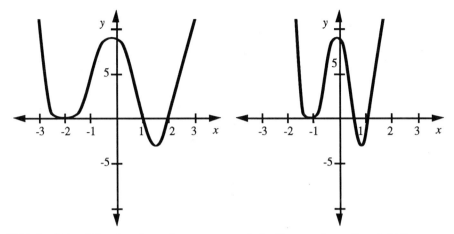

Which of the following formulas corresponds to the graph on the right?

$$y = f(2x) \qquad y = f(x+1) \qquad y = f(|x|)$$
$$y = f\left(\frac{1}{2}x\right) \qquad y = f(x-1) \qquad y = f(-|x|)$$
$$y = 2f(x) \qquad y = f(x)+1 \qquad y = f(-x)$$
$$y = \frac{1}{2}f(x) \qquad y = f(x)-1 \qquad y = -f(x)$$
$$y = |f(x)|$$

6(f). The three graphs correspond to a formula of the form $y = a(x-d)^2 + e$. Which of the parameters a, d, e are identical for the three given graphs?

7(e). Can the given graph correspond to a 3rd degree polynomial, a 4th degree polynomial, or neither?

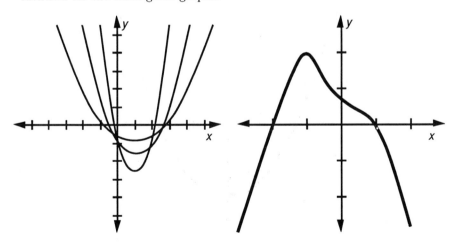

FIGURE 1.   Sample questions from the questionnaire.

1. Given the function $y = |x^2 - 4|$
(a) Graph the given function
(b) Graph $y = |x^2 - 4| + 1$
(c) Graph $y = |(x - 1)^2 - 4|$
(d) Graph $y = \frac{1}{2}|x^2 - 4|$

2. Given the function $f(x) = x^3 - 3x^2$ define $g(x) = f(x + 3)$. Find $g(-2)$.

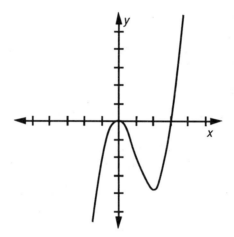

3. Given the graph of the function $y = f(x)$ sketch the graph of the function $y = \dfrac{1}{f(x)}$

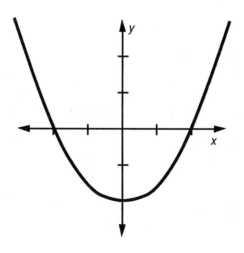

FIGURE 2.   Interview questions.

2.4 Procedure.

2.4.A The sessions. Because of the number of computers available to us in the school, two groups each consisting of four pairs of students, were formed. Each group met for six lessons spread over a five week period using the Green Globs software. The groups were taught in a very similar manner. The students attended, on the average, 4.8 out of the six lessons, and thus they had about 5 full hours of computer time to experiment with Green Globs.

We were with the students at all times during the lessons and acted as resource individuals. That is to say, we answered questions which were posed to us often by hints and questions rather than by direct response. We wanted the students to construct the bridges between the visual representation of a transformation and its algebraic description by themselves. Thus we never took the initiative to make suggestions to steer the students into new ways of thinking when they did not somehow arrive at the new ideas by themselves. A typical dialogue between a student (St) and us, the experimenters (Ex) was:

St: How can I make the graph of $y = \frac{1}{2}x^2 + 2x + 3$ wider?

Ex: Which coefficient in your expression could be responsible for width?

St: 1/2? Change 1/2?

Ex: Try it.

Another possible answer to the first question would have been: "What do you think would happen if you looked at $y = f(x/2)$?" The modus operandi guiding us in answering questions was to encourage the students to build the connections between the analytic and the graphical description themselves.

Part of each lesson was devoted to a guided discussion. During these sessions, attention was drawn to certain classes of functions and transformations to which the functions could be subjected. Specifically, the topics were linear functions in the first lesson, quadratic functions in the second lesson, the introduction of absolute values on the dependent and independent variable for linear and quadratic functions in the third and fourth lessons, and zeros and shifts of functions in general and polynomials in particular in the fifth and sixth lessons. During all lessons the role of the parameters in the expressions was stressed and discussed and summary statements were made. However, careful attention was paid not to direct the students to the test questions per se. The periods of instruction were short, usually taking about 10 minutes per session, in the middle or towards the end of the session.

Detailed notes on the students' actions were taken during all sessions. These notes described the types of functions used by the students, their rates of success in traversing globs, and the actions they took in order to improve a function which hit fewer globs than they had expected.

2.4.B Testing/Interviews. The main source of data came from a posttest and student interviews. The posttest was identical to the pretest (Section 3.3.B) and was given one week after the last lesson. The advantage of giving an identical test with its item by item comparability was considered to outweigh the disadvantage, namely the possibility of learning from the pretest. The results confirmed this expectation. In fact, if anything had been learned on the pretest, it would presumably have concerned questions of a type with which the students were familiar from the outset. But on the standard questions little progress was observed.

The interviews were held a few days later. One student from each pair was invited for an interview. Seven were actually interviewed, one by one, each for approximately half an hour. Each student was asked the same three questions (see Figure 2). The students were asked to think aloud as they started to do the exercises. Both experimenters were present at the interviews. One asked the questions and possibly gave hints while the other one took detailed notes of all interesting occurrences, including non-verbal actions and interactions. In addition, all interviews were recorded on audio-tape.

Question 1 showed whether or not the students made a connection between the graphs in b, c, d and the graph in a; in particular, whether they conceived of the latter three graphs as (slide/stretch) transformations of the first one. At every stage they were required to explain how they had constructed their graphs. This was relevant already for Part a (graph $y = |x^2 - 4|$) where we wanted to know whether they had plotted the parabola $y = x^2 - 4$ and flipped the lower part over the x-axis or whether they had obtained the graph by plotting points.

Question 2 was novel to the students; it was of interest to see how they viewed the new function. Were they able to see it as a transformed version of the old one (slide three units to the left) or did they look at it as a composite function? If they could not make sense out of the symbols in front of them, it was explained to them that the function g was defined as follows: The value of g at x equals the value of f at $x + 3$. When students started plugging numbers into the formula $g(x) = (x+3)^3 - 3(x+3)^2$, they were asked whether they could use the graph; if this did not help they were asked whether they were able to graph $y = g(x)$, and whether their formula for $g(x)$ suggested how to graph it. Students who gave the correct answer as $g(-2) = f(1) = -2$, were asked to explain graphically what it meant.

Question 3 showed to what extent they could analyze the graph of a function. They had never seen this type of question before but it was included to see if the teaching unit helped them develop the mathematical sophistication for processing functions graphically even beyond what they were taught. We wanted to determine if the students had internalized the notions of working with graphs so as to be able to construct new graphs from ones given without algebraic description. Theoretically, given the graph of f one can qualitatively construct the graphs of $f'(x)$, $\dfrac{1}{f(x)}$, $f^{-1}(x)$, $(f(x))^2$ etc., and in so doing, one demonstrates a firm understanding of the ideas underlying the transformations. Indeed, one demonstrates a certain amount of function sense.

3. FINDINGS AND INTERPRETATIONS

3.1 Test results. Figure 3 presents the pre- and posttest results, for both parts of the questionnaire. Scores increased on all non-standard questions (see Table 1). Particular note should be taken of Questions 4 and 6. These indicate that transformations on the parabola were tied to their algebraic representations. This idea however was not transferred to graphs in general, as evidenced by the results for Question 5.

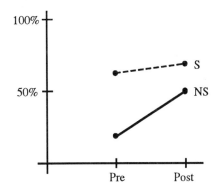

FIGURE 3.   Pre- and posttest results for standard (S) and non-standard (NS) questions (in percent).

TABLE 1. Pre- and Posttest Mean Scores on
Non-standard Questions (percent correct).

Question	3	4	5	6	7
(No. of parts)	(10)	(3)	(5)	(6)	(10)
Pretest	32%	27%	18%	29%	21%
Posttest	48%	56%	31%	76%	43%

3.2 Interview Results.

3.2.A Question 1. Question 1 revealed an interesting relationship between students' ability to find the correct answer and their method of solution.

(i) All seven students recognized Question 1(b) as a transformation but only 4 of them gave a correct answer.

(ii) Only four of the students recognized Question 1(c) as a transformation, and each of them, plus an additional student, gave a correct answer.

(iii) On Question 1(d) only two recognized the transformation, but all seven gave the correct answer.

The results of Question 1 are inconclusive; recognizing the transformation did not imply that one would give a correct result and vice versa, giving a correct result did not imply that one had recognized the transformation. Thus neither the answers to 1(b), nor to 1(d) can give any information about a relationship between recognizing the transformation and giving the correct answer (although a statistical computation would obviously give a strongly negative correlation). On the other hand, in Question 1(c), the transformation was recognized by 4 of the 7 students; they all gave a correct answer, while only one of the remaining three gave the correct answer. Although this evidence is weak, i t seems to indicate that recognizing the transformation did help them correctly answer the question.

3.2.B Question 2. Most students needed help with the formulation of Question 2, and even then only three were able to solve the problem to any acceptable depth. Out of the seven students, two took a graphical approach, two an analytic one, two were completely lost and one gave a correct answer based on his intuition, but was absolutely unable to give us any indication of how he had reached this conclusion. Of the two who had taken a graphical approach, one solved the problem completely using the transformation and gave a full explanation. The fact that he solved the problem by translation expressed itself also in that he got the final numerical result as "approximately negative 2.5" from the graph. The other one did mention shifting the graph but was unable to carry out the operation and reach any conclusion. The two who took an analytic approach solved the problem correctly by computing $g(-2)$ as $g(-2) = f(-2+3) = f(1)$, but they did not see any connection to a shift transformation. When asked whether they could have used a graph to solve the problem they both started to work on graphing the function g using $g(x) = (x+3)^3 - 3(x+3)^2 = x^3 + 6x^2 + 9x$ which they had obtained by algebra. In other words, for these students the function first had an algebraic description and, of secondary importance, this algebraic representation could be described graphically too. Computing $f(1)$ through direct substitution is the quickest way to solve this problem, but this way does not emphasize the connection between the algebraic and graphical representations of the function. Only one of the interviewed students made this connection fully and successfully. The others were not able to make this connection, even when asked explicitly.

3.2.C Question 3. This question was beyond the students in the study. Whatever they drew looked either like parts of parabolas, or like parts of straight lines, with no intelligible connection to the given graph.

3.3 Interpretations.

3.3.A Visual Processing. Because we consider visual processing to be one of the main components of function sense its importance can not be overemphasized. When students see $\sin(x)$ or $\log|x|$ or e^x it is also hoped that they will simultaneously picture a curve, and that this visual interpretation will be just as dominant for them as the analytic formulation. Students should have the graphs of certain "stock functions" in their repertoire (Lowenthal & Vandeputte, [**17**]) and they should feel at ease in manipulating them.

As a result of the program and our instruction there seemed to be more of a readiness among the students to approach problems by visual means, to use qualitative arguments, and to process information visually. In 24 out of the 42 answers (the seven students had to make a total of 6 responses each, four on Question 1 and one on each of Questions 2 and 3), the students used visual arguments in their solutions. Specifically, this was so for all students interviewed on Questions 1b and 1d, for about half of them on Questions 1a, 1c and 2, but for none of them on Question 3 . Thus a question naturally arises: Does visual processing contribute to the correctness of the answers? Seventy one percent of the problems processed visually were answered correctly. For problems processed non-visually, 22% were correct.

This seems to support the above claim concerning the importance of visual processing. Yet, it also raises the question why visual processing was not used more frequently by the students in the interviews. We suggest that working with the stock functions should form the basis for visually processing transformations on them. This means, for instance, graphically adding or multiplying functions such as f and g given by $f(x) = x$ and $g(x) = \sin(x)$ to $x + \sin(x)$ or $x\sin(x)$ by gestalting their graphs rather than adding or multiplying their ordinates . It is only when visual notions of the stock functions are internalized to the extent that they are at least as strong as their analytic formulations that one can visually perform transformations on them. While the experiment was not originally set up to check this claim, there are clear indications for it: The increase in students' success with transformations on parabolas (Question 6 of the questionnaire) as well as the frequency of their use of visual processing in the first interview question, points to the fact that they successfully used visual processing on those functions whose visual representation they had internalized, namely quadratics. On the other hand, even after instruction, they were far less successful on questions dealing with higher order polynomials.

3.3.B Transformation View. In some cases, the students did not consider the exercises as transformation problems. As an example, in Question 1(c), the students built the graph of $|(x - 1)^2 - 4|$ by first finding its vertex. They did not realize that this graph is just a shifted version of the graph they had constructed in part (a), not even with prompting. Indeed, this shows that in this case, they had no view of a function transformation at all. Similarly, numbers, independent of where they appeared, were viewed as entities instead of being part of the transformation. The -1 in the expression $|(x-1)^2 - 4|$ was not seen as determining an operation of shifting, but as just a number. In other words, the transformed stock functions were being viewed as completely new functions, unrelated to the stock functions from whence they came. On the other hand it can be assumed from the responses to Question 6 of the questionnaire and to Question 1 of the interview that in the simpler cases the transformed functions were viewed as results of transformations. Hence, it is natural to ask what characterizes a simple transformation so that it will be viewed as the transformation of a stock function and not as a completely new function? Two factors may enter here: Complexity of the function and complexity of the transformation.

In Subsection A we addressed the complexity of the functions. We now consider the complexity of the transformation. Although we have no hard evidence, it seems as though transformations in the vertical direction were easier for the students than those in the horizontal direction. In both of these cases, the shift/stretch transformations may be viewed as a composition of a given (stock) function with a linear function. For example, if f is the given stock function and t the linear function $t(x) = ax + b$, then $f \circ t$ is a horizontal shift/stretch, while $t \circ f$ is a vertical shift/stretch. There is a strong psychological difference between these two cases, with the horizontal one being conceptually more complex. In the horizontal case the composition appears explicitly, $f(x) \to f(ax+b)$, while in the vertical case it appears as an algebraic operation on the values of $f(x) : f(x) \to af(x) + b$. In a related study, Ayres, Davis, Dubinsky & Lewin [2] have investigated the relative complexity of the composition of two functions when either the first or the second is unknown.

On the basis of our observations of the students in the working sessions, we believe that the transformations used in this study are hierarchically ordered with respect to difficulty in understanding and mastering. For example, transformations which represent translation or stretching of a graph in a vertical direction seem to be more easily understood than similar transformations in a horizontal direction. This may be due to the complexity of the statements themselves, for much more is involved in visually processing the transformation of f to $f(x+k)$ than in visually processing the transformation to $f(x) + k$.

3.4 Additional Observations.

3.4.A Static versus Dynamic. Question 1 was the most suitable for assessing the static versus dynamic nature of students' views of function transformations. For example, do students view the transformation as *moving a graph from an initial state to a final one* with the graph having moved and changed throughout a transformation—or do they view it as a mapping which is moving every point in the plane to a new location? (See Edwards, [11], for details on this distinction.) The four parts of Question 1 were seen by all but one student as four separate problems; each such problem, for them, relates to two graphs, the initial and the final one, whose relation was described by the students: *The graph in b is higher than that in a, rather than: The graph in b can be obtained from that in a by moving it up.* Or: *The graph in d is wider than that in a, rather than: The graph in d is obtained from the one in a by reducing the height of each point to half its original value.* There was no view of the process, no view of a transformation performing some change. In other words, they viewed the transformations as a sequence of two static states rather than as a dynamic process. A similar phenomenon occurs at a simpler level with students who were asked to straighten out a wire by pulling at its ends (Piaget, [21]). As long as only traditional means (paper and pencil or blackboard) and computer graphing software such as Green Globs are used, students have no means to actually see the continuous transformation developing before their eyes; they must therefore conclude on the dynamic aspect from being given only the initial and final state. Movable transparencies, movie films, or computer software implementing transformations dynamically are apt to improve this situation. Such means were, however, not used in our teaching unit.

3.4.B Linearity Boundedness. Among some students there was a tendency to draw straight line graphs, regardless of the corresponding equation. In Question 1, two of the seven interviewees drew the required parabolic function $|x^2 - 4|$ by means of a partially linear graph, similar to the graph of $|x|+4$ or $|x-2|$. It is of interest to note that both then proceeded and solved two out of the next three parts of Question 1 correctly by applying a transformation to the graph. Similarly, in Questions 2 and 3 some of the students produced straight line bits where there were no indications of any linear properties. Similar linearity boundedness has been observed in other studies by Markovits, Eylon and Bruckheimer [18] among junior high school students, by Karplus [16] and Ponte [22] among high school students and by Dreyfus and Eisenberg [12] among college students.

3.4.C Super-polynomials. During the fifth session it was suggested that the students experiment with other functions, beyond linear, quadratic and absolute value functions and their composites. In particular, polynomials with specific zeros were suggested. Here, something interesting happened. Two pairs of students soon discovered that a large number of zeros caused the graphs to have large slopes when traversing these zeros. Such graphs look rather like a set of vertical lines, one at each zero. Using this and the fact that the globs are areas rather than points, it becomes easy to hit large numbers of globs with a single graph; such polynomials were called super-polynomials by Dugdale [9]. For example, the polynomial $(x + 7)(x + 4)(x + 1)(x - 1)$ will hit globs at $(-7, 2)$, $(-4, -4)$, $(-1, 3)$, $(-1, -3)$, and $(1, 8)$. The students realized how to use such polynomials to raise their scores on the game; but the underlying reasons why they worked seemed to evade them. They seemed not to realize that theoretical lines symbolize non-vertical curves with large slopes; they may not have known that the graphs continued outside the screen, *curved around* and *came back*, i.e. that all parts of the graphs were connected. In a game presenting globs at $(-9, -1)$, $(-8, -4)$, $(-6, 0)$, $(-4, -5)$, $(-1, -2)$, $(-1, 5)$, $(1, 4)$, $(3, -7)$, $(3, -1)$, $(5, -5)$, $(5, 3)$, $(7, 3)$, and $(8, 7)$ they would typically start with the polynomial

$y = (x - 3)(x - 5)(x + 1)$ and hope to hit six globs (see Figure 4). They did not understand why the graph in the example did not hit the two globs at $(3, -7)$ and $(5, -5)$.

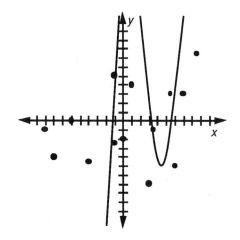

FIGURE 4.   A super-polynomial.

4. WHAT WE LEARNED: COMMENTS, HINDSIGHT & THEORIES

This teaching experiment was undertaken with no particular theory of learning in mind. The motivation for doing the experiment was based on the following reflections: A well instilled and developed function sense should be the central aim of the high school and the beginning collegiate curriculum. Function transformations form an important part of such function sense in view of both the calculus/analysis and the (modern) algebra components of the typical college curriculum. Visual reasoning ability should be an essential component of a high school graduate's mathematical skills. Visual reasoning can be taught. Function transformations are one topic where visual reasoning is particularly natural. The topic (function transformations) and the method (visual reasoning) are not dealt with in most standard curricula. Thus little is known about the opportunities and difficulties involved in teaching function transformations vis-a-vis a predominantly visual approach.

4.1 What we learned. There seems to be two broad categories of reasons as to why the students made limited progress during the teaching experiment. One of these categories arises because of difficulties with the content itself, function transformations; the other arises because of difficulties with the instructional format we used–but these difficulties are products of hindsight.

4.1.A Instructional format. The Green Globs program, in spite of the great press and accolades it has received, has its drawbacks. One of these drawbacks is that it is structured by being a game; students (and teachers) are finding themselves concentrating on how to increase their score by any means admitted by the software; this is a definite advantage at a moment in the curriculum when maximum flexibility in thinking about functions and their graphs is the aim; it is an equally definite disadvantage when specific aspects of functions and their graphs are the focus of attention. During the activities with Green Globs, our students rarely had an incentive to focus on transformations or even to view two graphs which they had drawn in sequence as related by a transformation.

For example, when graphing $x^2 + 4$ and then $x^2 - 2x + 4$, they may well have constructed the second graph on the basis of their knowledge about parabolas (vertex at $x = 1$, $y = 3$, etc.) rather than as a shift of the first graph by one unit to the right and by one unit down: $[(x-1)^2 + 4] - 1 = x^2 - 2x + 4$. The connection of the students' activities with transformations was established in the guided discussion sessions that followed the activities. But as many teachers well know, it is not always easy to switch students' attention from their computer screens to the group discussion. Moreover, the sequence of discussions was structured, at least initially, according to the type of function rather than according to the type of transformation: linear functions in the first meeting, quadratic ones in the second. As a consequence the absolute value functions introduced in the third and fourth meetings may have been conceived by the students as a new, more or less independent class of functions rather than variants of linear and quadratic functions. Therefore, the same transformations, which were omnipresent for the instructors during the activity sessions, could be noticed, comprehended and later actively used only by those students who were able and willing to follow the instructors when they made the connections across various meetings during the group discussion.

Finally, Green Globs is made for graphing single functions whereas transformations relate at least two (initial and final state) or an infinity (an entire set of intermediate states) of functions and their graphs. Thus, viewed from today's perspective when more powerful programs are available (see for example Kaljumagi, [15]), Green Globs does not provide maximal support for learning activities about function transformations, nor did we use it in the most effective way because we did not establish a smoothly guided learning sequence to build the mental structures to understand the effect of general transformations. Superpolynomials are one example of a detractor which was not planned by us but anyway suggested by some of the students, which is an eminently suitable and clever idea for use in Green Globs, teaches some valuable knowledge about functions but has little to do with function transformations. The learning sequence we established was, from the point of view of function transformations, more like a roller-coaster, with transformations showing up more or less at random as dictated by the aims of the game (get a high score).

4.1.B Concepts and approach. Learning about function transformations is difficult. This is not surprising: it presupposes a good understanding of functions, at least in their algebraic and graphical representations and of the relationship between them; it moreover requires establishing links between two or more such functions, each one a structure including graph, algebra and links between them. From the test and interview results it can be seen that most students in our experiment acquired a good grasp of what it means to transform $f(x) = x^2$ into $g(x) = a(x-d)^2 + e$ including the effects of the signs and magnitudes of the coefficients a, d, and e. However, most of them did not transform other functions at all. They seemed to look at each transformed function as an independent function. In other words, they seemed unable to transfer this knowledge to higher order polynomials and more general classes of functions; we must assume that they had not built a general mental image for what a transformation such as a shift or stretch does to a function (including its algebraic and graphical representations) although they were able to correctly apply such transformations to quadratics.

Not only were the students more successful in dealing with transformations on some functions and less on others but we also observed that certain classes of transformations seemed to be easier for students than others. In particular, transformations in a vertical direction ($f(x) \to f(x) \pm k$, $f(x) \to |f(x)|$, $f(x) \to \pm k f(x)$) were easier for students than the corresponding horizontal direction ($f(x) \to f(x \pm k)$, $f(x) \to f(\pm kx)$, $f(x) \to f(|x|)$). A more detailed hierarchy of transformations, given in Figure 5, merits further attention. At the very least Figure 5 can provide a basis for making a systematic inquiry into this observation.

4.2 Theories. The findings and interpretations of this study can be discussed in a theoretical context. Several theories have been proposed on how students acquire notions of functions, and various aspects of these theories do shed light on several of our observations. Some of these theories of how students acquire knowledge of functions overlap with one another. Nevertheless, we apply appropriate aspects of them to some of our observations below.

4.2.A The reflective abstraction chain. Douady [6], Dubinsky [8], Sfard [27] and Thompson [29] have all proposed theoretical frameworks for the description of learning processes in which mathematical knowledge is characterized in terms of mathematical objects and processes which transform these objects. Breidenbach, Dubinsky, Hawks and Nichols [3] and Sfard [28]) have both applied their respective theory to learning the function concept. A function may be considered as a process that associates objects from the codomain with objects from the domain. Often these objects are numbers (in all examples in this paper, they are in fact numbers). Any specific function may be considered as a process that operates on these numbers: it transforms them into other numbers. For many purposes, it is, however, inadequate to consider a function as a process; for example differentiation is a process that operates on and transforms functions. Composition of functions is another such process (Ayres, Davis, Dubinsky

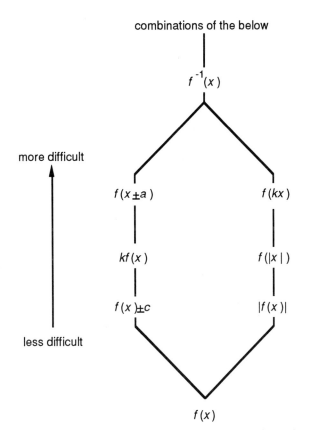

combinations of the below

$f^{-1}(x)$

more difficult

$f(x \pm a)$ $f(kx)$

$kf(x)$ $f(|x|)$

$f(x) \pm c$ $|f(x)|$

less difficult

$f(x)$

FIGURE 5.   A hierarchy for function transformations.

& Lewin, [2]); so are the transformations dealt with in this paper. For understanding them, it becomes necessary to consider the function itself as an object,
on which a transformation acts. We have thus climbed up one level and made the
process of the lower level into the object of the higher level. (This description,
in fact, covers two levels out of a much longer sequence of levels, but these are
the two levels which are relevant for this paper.)

One reason for the complexity of mathematical knowledge is that most mathematical notions may take the role of processes or objects, depending on the
problem situation and the student's conceptualization. Dubinsky [8] has developed a theory to describe the learning process toward a process-object conception
of function (and other notions). Just as Thompson's, his ideas are based on and
extend Piaget's, mainly the notion of reflective abstraction. The application
of this theory to the function concept is described in Breidenbach, Dubinsky,
Hawks & Nichols [3]. Following them, an action is a repeatable manipulation
of objects; in terms of functions, the plugging of numbers into an expression to
find the value of the function would be such an action. An action conception of

function is static whereas a process conception is a dynamic view of the transformation of objects (numbers in the case of the type of function used in our paper). An object conception of function is not described in their paper but it follows from what has been stated above that the central feature of an object conception of function is the ability to conceive of a function as something on which an action such as differentiation, composition or a transformation can be performed. The action/process/object view of functions can be extended in a natural manner to an action/process/object view of transformations. In fact, the extension is so straightforward that we will not spell it out here.

It follows from the theoretical view expounded above that students are unlikely to achieve even a partial understanding of the complex structure of the idea of transforming a function unless they have some grasp of a function as an object. Similarly, it has been pointed out by Selden and Selden [26] that every calculus student should ideally treat functions as objects when acting on them by differentiation and integration but that in reality many fail to realize that they are dealing with functions as objects because they carry out these operations automatically.

Let us draw the parallel to function transformations. As our informal experience from teaching beginning college courses shows, most students can learn to answer questions of the type "What is the (algebraic rule of the) function obtained by shifting $f(x) = \sin(\pi x)$ to the right by 1.5 units." They learn that "shifting" requires them to "substitute $x \pm a$ for $x$rq' , "to the right" requires them to "use the minus sign" and "by 1.5" requires them to insert 1.5 for a; thus they write down $\sin(\pi(x - 1.5))$ and some might even write this as $\cos(\pi x)$. The word "shift" acts for them as a cue in a similar manner as elementary students use the word "less" as a cue for subtraction in word problems - whatever the accompanying text. Similarly, they learn to deal with other shift and stretch transformations, most of which are easier than the given example because they contain fewer opportunities for mistakes such as wrong signs or parentheses (e.g., writing $\sin(\pi x - 1.5)$ instead of $\sin(\pi(x - 1.5))$). It is possible to carry out such substitutions in a mechanistic manner – and all they require is a good memory for things such as whether "to the right" is the prompt for the plus or the minus sign.

With respect to our teaching experiment, it is a moot point whether such students should be characterized as having an action conception of transformation or not, because we purposely avoided opportunities for such a mechanistic approach and execution of function transformations. Our students may have been faced with a situation where they wanted to get a graph "like the one they had but further to the right;" they may even have had rather precise information about how much further, say 1.5 units. But they were not cued by an abstract, possibly for them meaningless, word like "shift;" they were faced with a concrete and purposeful need to move that graph; they may or may not have associated the word "shift" with this operation; the point is that they were put into a situation in which meaning was necessarily associated with the operations they were about to carry out.

Why, then, did many of our students succeed only partially on the posttest and the interview? Obviously, they had not yet successfully abstracted the idea of what it means to subject a general function to a transformation. They certainly had not constructed a fully elaborated process conception of a transformation - otherwise they would have succeeded much better in answering Question 5 of the posttest and Question 2 of the interview. On the other hand, the results of interview Question 1 allow the assumption that about half of the students did have a process conception of transformation, at least as long as the functions they had to transform were quadratics. The results of posttest Question 4, and more clearly Question 6, lend support to that assumption.

We had not attempted to set up our questions so as to identify how far on the way towards a process conception of transformation a student stood. Therefore the above assumptions must remain assumptions; there is no experimental evidence to confirm or contradict them. The lack of appropriate data makes a detailed analysis of a single student's conceptions of transformations impractical. We therefore content ourselves with suggesting two possible reasons for the limited success of the teaching experiment , in addition to the reasons related to the instructional format which were already stated in 4.1.A. These two reasons are of a rather general nature, based on the above theoretical considerations.

4.2.B Transformations as static versus dynamic. We studied how students acquire the notion of function transformation. Underlying our study was a dynamic view of a transformation as mapping the Cartesian plane into itself. By studying the original position of the function and its new (transformed) position, students should be able to build a series of transformations in their mind as to how it got there. However, when the students focused on the old state – new state situation, they were considering a transformation as something static; they were considering the product of the transformation only: the resulting graph. If they saw a series of transformations which literally was sliding stretching, shrinking, twisting and turning the old graph into a new one (i.e., moving every point to a new location through a series of intermediate states), then they were using a process model of the notion. Understanding both models is important to have a full grasp of what it means to subject a function to a transformation. The only model of transformation which the students in this experiment seemed to have was that of a transformation as a product.

4.2.C Object conception of function. It can be argued that students' success was limited because their conception of function itself (rather than transformation as in the previous paragraph) was only a process conception (or less) rather than an object conception. If the students did not have an appropriate understanding of the notion of a function from the outset, this experiment could not have succeeded no matter what the circumstances of instruction. We did not check to see at the start of the experiment to what extent the students understood the notion of a function. With hindsight, this was an oversight on our part, and this should have been checked even with this mathematically advanced population. It should, however, be remarked here that there is a certain inconsistency between this argument and the fact that students succeeded well with quadratic functions. Could it be that they had an object conception of quadratic functions but not of others? Are process and object conceptions of function transformation acquired progressively, first for better known functions (quadratics) and later for other classes of functions? More generally, the growth of a process conception not only of function transformations but of any mathematical notion might depend rather strongly on the kind of (lower level) objects with which the process deals, on which it operates. Whether a student, in a particular problem situation, acts according to a process conception of a certain notion or not might depend on many other factors such as the complexity of the problem situation. Even if theories such as the ones referred to above give us a general framework for analysis, much work has to be done to adapt them to the analysis of student behavior in particular situations.

5. SUMMARY

Delineating facets of mathematical thinking in general and function sense in particular seems to be a worthwhile activity. This paper contributes to our understanding of how difficult it is to master the notion of a function transformation. Underlying reasons for this difficulty are cited in Section 4 and this listing will hopefully be used in guiding further studies. Like many teaching experiments, this study raised more questions than it answered. But developing a sense for functions, and in particular for function transformations is, in our opinion, an extremely important but difficult task to achieve. One aspect of understanding the notion of transformations is to visualize their effects; we believe that such visualizations can be taught. This study was one small step to help us understand how students learn this skill.

References

1. *Focus issue: number sense.*, Arithmetic Teacher **7** (1989).
2. Ayres, T., G. Davis, E. Dubinsky & G. Lewin, *Computer experiences in learning composition of functions*, Journal for Research in Mathematics Education **19(3)** (1988), 246–259.
3. Breidenbach, D., E. Dubinsky, J. Hawks & D. Nichols, *Development of the process conception of function*, Educational Studies in Mathematics **23** (1992), 247–285.
4. Clements, M. A., *Terence Tao*, Educational Studies in Mathematics **15** (1984), 213–238.
5. Davis, F. E., *Green Globs gets good grades.*, A+ The #1 Apple II Magazine **5(10)** (1987), 22.
6. Douady, R., *The interplay between different settings: Tool-object dialectic in the extension of mathematical ability - examples from elementary school teaching*, Streefland (ed.) Proceedings of the ninth international conference for the Psychology of Mathematics Education, Vol. 2: Plenary Addresses and Invited Papers (1985), 33–52 Noordwijkerhout, The Netherlands.
7. Dreyfus, T., & T. Eisenberg, *The function concept in college students: linearity, smoothness and periodicity*, Focus on Learning Problems in Mathematics **5(3/4)** (1984), 119–132.

8. Dubinsky, E., *Reflective abstraction in advanced mathematical thinking*, D. Tall (ed.) Advanced Mathematical Thinking, Kluwer, Dordrecht, The Netherlands, 1992, pp. 95–123.

9. Dugdale, S., *Green Globs: a microcomputer application for graphing of equations*, Mathematics Teacher **75(3)** (1982), 208–214.

10. Dugdale, S. & D. Kibbey., *Graphing equations (Software)* (1982), Iowa City: Conduit.

11. Edwards, L., *The role of microworlds in the construction of conceptual entities*, G. Booker, P. Cobb & T. N. di Mendicuti (eds.) Proceedings of the fourteenth international conference for the Psychology of Mathematics Education, vol. 2, Mexico, 1990, pp. 235–242.

12. Eisenberg, T. & T. Dreyfus, *On the Reluctance to Visualize in Mathematics*, W. Zimmerman and S. Cunningham (eds.) Visualization in Teaching and Learning Mathematics, vol. 19, USA Mathematical Association of America. Notes Series, 1991, pp. 25–37.

13. *Everybody Counts* (1989), National Academy Press, Washington, DC..

14. Greeno, J. G., *Number sense as situated knowing in a conceptual domain*, Journal for Research in Mathematics Education **22(3)** (1991), 170–218.

15. Kaljumagi, E. A., *A teacher's exploration of personal computer animation for the mathematics classroom*, Journal of Computers in Mathematics and Science Teaching **11 (3/4)** (1992), 359–376.

16. Karplus, R., *Continuous functions: students' viewpoints*, European Journal of Science Education **1** (1979), 379–415.

17. Lowenthal, F. & C. Vandeputte, *Manipulations of Cartesian graphs: A first introduction to analysis*, Focus on Learning Problems in Mathematics **11(1/2)** (1989), 89–98.

18. Markovits, Z., B. Eylon & M. Bruckheimer, *Functions today and yesterday*, For the Learning of Mathematics **6(2)** (1986), 18–24.

19. Mason, J., L. Burton & K. Stacey, *Thinking Mathematically*, Addison Wesley, London, UK, 1982.

20. Mundy, J., *Analysis of errors of first year calculus students*, A. Bell, B. Low, & J. Kilpatrick (eds.) Theory, Research and Practice in Mathematics Education. Fifth International Conference on Mathematics Education, Adelaide, 1984: Working group reports and collected papers, Nottingham, UK: Shell Center, 1985, pp. 170–172.

21. Piaget, J., *Experimental Psychology, its Scope and Method*, Routledge, London, UK, 1968.

22. Ponte, J. P. M., *Functional reasoning and the interpretation of Cartesian graphs (Doctoral Dissertation, University of Georgia , Athens)*, Dissertation Abstracts International, vol. 45(06), 1675-A. (University Microfilms No. 8421144), 1984.

23. Presmeg, N., *Visualization and mathematical giftedness*, Educational Studies in Mathematics **17** (1986), 297–311.

24. Romberg, T. A., *Perspectives on scholarship and research methods*, Grouws, D. (ed.) Handbook of Research on Mathematics Teaching and Learning, Macmillan, New York, NY, USA, 1992, pp. 49–64.

25. Schwarz, B., *The Use of a Microworld to Improve Ninth Graders' Concept Image of a Function: the Triple Representation Model Curriculum* (1989), Doctoral Dissertation, Weizmann Institute of Science, Rehovot, Israel.

26. Selden, A. & J. Selden, *Research perspectives on conceptions of function: summary and overview*, G. Harel & E. Dubinsky (eds.) The Concept of Function: Aspects of Epistemology and Pedagogy, vol. 25, Mathematical Association of America, Notes Series, USA, 1992, pp. 1–16.

27. Sfard, A., *Two conceptions of mathematical notions: Operational and structural*, J.C. Bergeron, N. Herscovics & C. Kieran (eds.) Proceedings of the eleventh international conference for the Psychology of Mathematics Education, vol. 3, Montreal, Canada, 1987, pp. 162–169.

28. Sfard, A., *Transition from operational to structural conception: the notion of function revisited*, G. Vergnaud, J. Rogalski & M. Artigue (eds.), Proceedings of the ninth international conference for the Psychology of Mathematics Education, vol. 3, Paris, France, 1989, pp. 151–158.

29. Thompson, P., *Experience, problem solving and learning mathematics: considerations in developing mathematics curricula*, E. Silver (ed.) Teaching and Learning Mathematical

Problem Solving: Multiple Research Perspectives, Lawrence Erlbaum, Hillsdale, NJ, USA, 1985, pp. 189-236.

30. Vinner, S., *Avoidance of visual considerations in calculus students*, Focus on Learning Problems in Mathematics **11(1/2)** (1989), 149–156.

BEN-GURION UNIVERSITY, BEER SHEVA, ISRAEL

CENTER FOR TECHNOLOGICAL EDUCATION, ISRAEL HOLON, ISRAEL

CBMS Issues in Mathematics Education
Volume **4**, 1994

Three Approaches to Undergraduate Calculus Instruction: Their Nature and Potential Impact on Students' Language Use and Sources of Conviction

SANDRA FRID

ABSTRACT. This report contains research results from task-based interviews with undergraduate, introductory calculus students. The analysis focuses on the students' language use and their source of conviction regarding how mathematical truth and validity are determined. Three undergraduate classes at three different institutions were taught using one of three instructional approaches: technique oriented, concepts first, and infinitesimal instruction. Each approach is described on the basis of systematic classroom observations. Three different types of students emerged in the results across all three classes. The three groups, Collectors, Technicians and Connectors, each differ in the degree to which their sources of conviction were internal vs. external. Analysis of students' language use indicated that they varied in use of pre-calculus and everyday language, visually and physically oriented language, and procedural language knowledge in their responses to calculus problems. Students who received infinitesimal instruction used more everyday language in their responses. Their language was also more integrated with symbols and technical language.

In recent years, educators have recommended a number of curriculum reforms in calculus instruction (for example, see [**7, 20, 33**]). Recommended changes include: (1) shifting the focus of calculus teaching to the fundamental ideas of calculus, rather than emphasizing drill in routine skills and techniques, (2) integrating applications into the body of calculus courses by reinforcing the role of approximations and problem situations with contexts relevant beyond the field of mathematics, (3) producing textbooks to support curriculum changes, and (4) integrating computing technology into calculus curricula. However, research into the teaching and learning of calculus is in an embryonic state. Studies are needed of various instructional emphases and formats and their subsequent effects on learning.

Although some educators who favour change in introductory calculus instruction have been developing and implementing changes (for example, see [7, 16, 33]), it has been said a crisis exists in the teaching of undergraduate calculus [3, 23, 24]. Student drop-out and failure rates in calculus remain high, between 30% and 50%, compared to many other undergraduate courses. Students passing a calculus course tend to perform at low levels with respect to both skills and the use of calculus ideas [3, 24]. Unfortunately many undergraduate students must succeed in an undergraduate calculus course either to begin or to continue enrollment in many programs of study. Calculus is a required course for students aiming for careers in science, engineering, medicine, business, education and other fields.

In the last decade researchers have frequently investigated student understandings of limits (for example, see Davis & Vinner, [6]; Sierpinska, [29]; Tall & Vinner, [34]; Williams, [37]). Student understandings of differentiation or integration have been investigated to a lesser extent. In addition, many studies focus largely on student errors, misconceptions, or inability to perform certain tasks (for example, see Selden et al., [28]; Davis & Vinner, [6]; Orton,[21, 22]; Williams, [37]). A number of studies related to student achievement in calculus have also been conducted (for example, see Edge & Friedberg, [8]; Hirsch et al., [17]; Selden et al., [28]), as well as a limited number of reports and discussions of the general state of calculus instruction (Cipra,[3]; Douglas,[7]; Peterson, [23, 24]). More recent research centers on the influences on student learning of non-traditional calculus curricula or teaching strategies (for example, Heid, [16]; Tall, [33]). Although changes have been occurring in the teaching of calculus, relatively little research literature related to student learning arising from these changes has appeared.

The findings of the study reported in this paper address the issue of the impact on student learning of alternative approaches to calculus instruction. The report examines the learning of undergraduate students taught by one of three approaches to calculus instruction. Two of the three approaches to calculus instruction which formed the research setting were developed by instructors at the related institutions to address high drop-out and failure rates in undergraduate calculus. These two post-secondary institutions used their own locally developed calculus curricula. The staff developed their course rationale, approach, and supporting textbook materials to fulfill some of the suggested changes in the academic literature already outlined.

More specifically, the study reported here was designed to address a need to investigate student learning in calculus from a perspective likely to reveal new insights into student learning and potential relationships between instruction and learning. The emergence in mathematics education research in the last decade of the theory of constructivism offered a perspective by which such an examination might be effectively implemented. Constructivism views mathematics learning as an active, constructive process in which individuals build knowledge for themselves (Schuell, [27]; von Glasersfeld, [35, 36]). This view contrasts the more traditional view of learning that sees concepts transferable "ready-made" from teachers to learners. In particular, constructivism views learning as an adaptive process which, through trial and error, an individual constructs a viable model of the world (von Glasersfeld, [35]). This model is seen by constructivism as a fit of knowledge to experience, so that instead of discovery of an "independent, pre-existing world" (Kilpatrick, [19]; p.7), learning is viewed as adaptation to what is experienced within the world. Prior knowledge plays a key role in this process, whereby individuals' existing conceptual structures and processes act to determine how they interpret and comprehend experience.

Constructivism emerged as an important influence in mathematics education research because it provided a valuable perspective from which to understand mathematics learning (Ernest, [9]). More recently, Ernest [10] examined constructivism through discussion of mathematics as a social construction. This philosophy, known as social constructivism, takes the view that "human language, rules, and agreement play a key role in establishing and justifying the truths of mathematics" (p.42). That is, mathematical knowledge is grounded in: (1) "linguistic knowledge, conventions and rules" (p.42), (2) social processes by which an individual's internal, subjective knowledge is turned into external, objective knowledge, and (3) objectivity viewed as public, social acceptance rather than an inherent property of the content of knowledge. These features imply mathematical knowledge depends on social sharing of language and decisions pertaining to truth and validity. Further, as a consequence for mathematics education researchers, these points imply language use and ways of determining truth and validity are likely to be important components of mathematics learning.

Thus, this study was designed to investigate student learning in calculus with a focus on language use and the ways truth and validity are determined. The term sources of conviction refers to how one determines truth and validity. More specifically, sources of conviction refer to how one determines facts, accordance with accepted mathematical principles and standards, legitimacy, consistency and logicality. The study examined the following:

1. the nature and role of the language students use to interpret calculus concepts and problems,

2. the nature and role of students' convictions regarding the validity or truth of calculus interpretations and problem responses,

3. the ways students construct their calculus conceptualizations,

4. the ways different approaches to calculus instruction impact on students' language use, sources of conviction and manner of construction of conceptualizations, and

5. the ways three different approaches to calculus instruction translate into classroom, textbook and exercise assignment instructional events.

RESEARCH METHODS

The research, a naturalistic study, involved three undergraduate calculus classes located at three different post-secondary institutions in Western Canada: a large university and two small private colleges. These institutions will be referred to as Alpha University, Beta College, and Gamma College. The corresponding instructors will be referred to as Professors Alpha, Beta , Gamma, respectively.

Interviews.

Seventeen students took part in task-based and personal interviews as methods of inquiry into their language use, sources of conviction, and manner of construction of conceptualizations. Ginsburg [13] outlines how clinical task-based interviews are appropriate for psychological research on mathematical learning because they allow discovery and identification of cognitive structures and thought processes. The interviews were conducted in the last three weeks of a 13 week school term. Each interview lasted one to two hours. A 20 to 30 minute follow-up interview was conducted two to three weeks later for the purpose of allowing each student to clarify or expand upon responses from the first interview. These in-depth interviews involved students in oral and written responses to a number of calculus problems focusing on calculus skills and concept interpretations (examples of these problems can be found in the section on language use, as well as in Appendix A). The problems included open-ended as well as relatively focused tasks which asked students to identify, describe, interpret, explain or apply limit and derivative concepts. The problem set was selected to provide a range of mathematical representations within which students could work, including words, symbols, graphs, and applications. It was also designed to provide opportunities for translation among these various forms of mathematical representation. The interviews provided extensive performance examples and detailed accounts of students working through calculus problems. The interviews also intended to probe students' beliefs about calculus and their calculus learning. Thus, the interviews incorporated relevant personal questions related to students' perceptions of calculus and the learning of calculus, study practices, ways of determining "correctness" and attitudes towards calculus.

Content and typological inductive analysis of each student's interview transcript and corresponding written responses determined students' sources of conviction, manner of construction of conceptualizations and language use (as outlined in Goetz & LeCompte, [14]). In particular, the overall manner of proceeding with analysis of students' sources of conviction and manner of construction of conceptualizations was similar to that used by Belenky et al. [1] to analyze interviews conducted "to explore with women their experience and problems as learners and knowers" (p.11). Particular attention was paid to students' statements and comments while responding to the calculus problems and to their references to strategies for learning calculus, personal sense of understanding of calculus, or specific experiences both inside and outside the calculus class-room. Salient features and quotations from each interview were then used to identify and describe categories of students' sources of conviction and manner of construction of conceptualizations.

Examination of students' language use was initially done on the broad level of an entire problem response. This involved counting the occurrences of a student's use of symbols and technical or everyday language that were not supplied in the problem statement. In relation to symbol use, manipulations or operations with symbols present in the statement of a problem were distinguished from a student's use of a symbolic representation not present in the statement of the problem. For example, it was noted if students merely performed operations with symbols in responding to a differentiation problem or if they introduced operator or first principles notation as a representation of the differentiation process. Use of symbols to label points, axes, or functions on a graph or diagram were distinguished from introduction of symbols that independently constituted a mathematical representation. Results of the counts of students' use of symbols, technical and everyday language terms were used to determine the nature and role of their language use, while the role of their language use was determined from more extensive examination of what they said or wrote and what this language reflected of their calculus conceptualizations.

Classroom Observations.

Systemic classroom observations and the use of low-inference variables, as outlined by Croll [5], were carried out for a 13 week school term from September to December. Each class was observed for 25% to 50% of regular classroom time. The low-inference variables monitored whether an instructor made truth and va-lidity claims by use of previously established mathematics or logical necessity, physical contexts (such as a graph), or rules that had not been previously derived or were not used with justification as to the choice of rule. Low-inference vari-ables were also used to monitor when an instructor used symbols, or technical or everyday mathematical terms or phrases. A distinction was made between spoken and written occurrences of these language variables. The observations thereby provided a quantitative description of the nature of truth and validity claims and language use in each instructional setting.

The researcher quantified instructional features related to the research ques-tions by observing classroom procedures. The range of instructional settings allowed partial examination of the impact of different approaches to instruc-tion on the nature and role of students' sources of conviction and language use, although the small number of students interviewed at each institution did not permit elaborate statistical analysis or definitive answers on the effect of instruc-tion upon students' calculus learning. The examination of the students' problem responses, however, gave insight into the potential impact of each instructional approach on all students' learning.

RESEARCH SETTING:

THREE APPROACHES TO UNDERGRADUATE CALCULUS INSTRUCTION

The following brief descriptions of the three instructional approaches provide outline the broad educational context of each approach in order to place the student interview data within the broader context of the instruction students received.

1. Introductory Calculus at Alpha University: Technique-Oriented Instruction.

Alpha University is a large urban university enrolling approximately 25,000 full-time students. The introductory calculus course at this university represents similar courses across North America. A standard textbook used for this course, Single Variable Calculus (J. Stewart, Brooks/Cole Publishing Company, 1987) covered content including limits, introduction to the derivative, derivative rules, applications of the derivative, the definite integral, techniques of integration, and applications of the integral. The instructional approach is "traditional" in that emphasis is put on learning techniques for differentiation, integration, graphing and the solution of textbook problems. Much class time and textbook space is devoted to methodically providing examples of such techniques. Generally briefly introduced intuitively or informally, then followed by precise definitions, related theorems and proofs, and numerous example problem solutions. The approach to instruction used at Alpha University will be referred to as "technique-oriented" instruction.

2. Introductory Calculus at Beta College: Concepts-First Instruction.

Beta College, a small college located in an agricultural region in a town of approximately 13,000 people, enrolls about 800 full-time students. Beta College uses what will be called a "concepts-first" approach to introductory calculus instruction. This explores concepts intuitively before introducing their formal definitions and proofs and before skill development is emphasized. This intuitive exploration uses the following approach: (1) experimentation with subcases of a concept, (2) examination of numerical, geometrical, or graphical representations of simple examples of a concept, and (3) comparison to concepts students have already been taught. Rules for specific skills are then presented, including demonstration of their plausibility. Lastly, concepts are revisited and developed in their precise, logically derived forms. For example, the derivative concept is developed in the following steps: (1) secant and tangent lines on simple functions are examined numerically and graphically, (2) comparisons are made between slopes of lines or functions and rates of change, (3) derivative rules are introduced, justified by demonstration of the differentiation process with a specific function and practiced with simple examples and (4) the precise definition of the derivative is given and related theorems proved rigorously.

The calculus instructors at Beta College developed their own textbook so it would match the concepts-first approach to instruction they wished to use. Thus, it differs from most calculus textbooks in the ordering of units. The book follows a spiral approach, with topics revisited more than once to be developed further or in different ways. However, the topics covered are the same as those in standard calculus courses in that they focus on limits, derivatives and integrals, and related skills and applications. At the time this research study was conducted, Beta College was beginning a second school year of use of the new textbook. Thus, the impact of the new curriculum on the withdrawal and failure rates in the course had not been determined.

3. Introductory Calculus at Gamma College: Infinitesimal Instruction.

Gamma College is a small urban college enrolling about 2000 full-time students. The introductory calculus course at this college uses what will be called an "infinitesimal" approach to calculus instruction. This approach develops concepts intuitively while using methods related to nonstandard analysis as analytic and computational tools. Infinitesimal methods are the tools by which Newton and Leibniz first developed calculus in the late 1600's. Newton and Leibniz did not precisely define infinitesimal numbers, demonstrate their algebraic properties, nor logically validate computations made with them. As a result, many mathematicians saw infinitesimal methods as a source of unsoundness in the classical foundations of calculus. Of course in the late 1800's Weierstrass and others rigorously founded calculus on real number concepts. Since then, real analysis methods have dominated the teaching of calculus.

In the 1960's Abraham Robinson used a logically rigorous approach to redevelop calculus in terms of infinitesimal number concepts (Robinson,[26]). Based upon mathematical logic, Robinson's treatise (called nonstandard analysis) is beyond the capabilities of most undergraduate students. However, his work can be translated to a form suitable for introductory calculus instruction by introducing students to infinitesimals intuitively. Then, with these numbers, calculus concepts are developed in both intuitive and formal ways.

The best way to demonstrate how this approach differs from the use of methods in real analysis, and in particular from technique-oriented and concepts-first approaches to instruction, is to provide some specific examples of its use. Instructors at Gamma College developed the following two examples of this approach:

1. Limits and their precise $\epsilon - \delta$ definition are replaced by the more intuitive notion of "rounding off" (an idea students have used since elementary school). In other words,

$$\lim_{x \to 0} \frac{x^2 + 2x}{3} = 0$$

is replaced by

$$f(\epsilon) = \frac{\epsilon^2 + 2\epsilon}{3} \rightsquigarrow 0$$

where if ϵ is infinitesimal then $\frac{\epsilon^2 + 2\epsilon}{3}$ is *very* small in size. Therefore, $\frac{\epsilon^2 + 2\epsilon}{3}$ is infinitesimal in size and will round off to zero.

2. The derivative is not introduced via rotating secants which in the limit become a tangent line at a point on a graph. Rather, the value of the derivative at a point is the slope of the tangent line at that point (if the tangent line exists). This concept of derivative is introduced after tangent lines (and where they do and do not exist) have been introduced via the intuitive notion of magnification, a process whereby a curve is magnified infinitely around a point. If the outcome of magnification looks like a straight line, this line is the tangent line. Both examples are based upon the intuitive notion of "close to." In infinitesimal instruction infinitesimal numbers are used for a more formal, mathematical justification of "close to," hence, we use the term "infinitesimal" to name this approach to instruction.

The textbook used for the infinitesimal approach to instruction was written specifically for use in calculus courses at Gamma College. This textbook was like the textbooks for the other two instructional approaches in that calculus was presented as a highly structured discipline where the ultimate goal is symbolic representation and justification. The book provides numerous definitions, examples, theorems, and proofs. It did however differ substantially in the approach taken for introduction and justification of concepts. This approach was done exclusively by infinitesimal methods.

RESULTS

Systemic Classroom Observations.

Results of the systemic classroom observations indicated instruction at the three institutions was similar in the extent of use physical contexts, unjustified rules for truth and validity claims, and the use of mathematical symbols, written or spoken technical language, and written everyday language. However, infinitesimal instruction as implemented by Professor Gamma displayed a greater use of spoken everyday language and use of previously established mathematics or logical necessity in truth and validity claims. The extent spoken everyday language used in the infinitesimal approach was 49% greater than that used for technique-oriented instruction and 33% greater than that used for concepts-first instruction. The extent of use in infinitesimal instruction of the structure of mathematics or logical necessity in truth and validity claims was 66% greater than that used for technique-oriented instruction and 37% greater than that used for concepts-first instruction.

Students at Gamma College used more everyday language than the other students. They also exhibited a higher degree of appropriate integration of everyday technical language and symbols. These features of Gamma College students' language use will be discussed in more detail in a later section, but note here Gamma College students' more extensive use of everyday language might be related to the greater extent of everyday language present in infinitesimal instruction as delivered by Professor Gamma.

Students' Sources of Conviction and Construction of Conceptualizations.

Student interview data revealed the existence of three groups of students who differed in their sources of conviction. These groups were named Collectors, Technicians, and Connectors. The names reflect the nature and role of the groups' sources of conviction. For the 17 interviewed students, eight were classified as Collectors, four as Technicians, and five as Connectors. Prominent features of each of the three groups are summarized and then discussed in the following sections. More complete accounts can be found in Frid [11]) and Frid [12].

Collectors.

Students classified as Collectors displayed sources of conviction that were generally external in nature; external in that they resided in statements, rules and procedures presented by the teacher or textbook. Generally, they did not reside in what students felt they had made sense of for themselves. The students, apparently, constructed their mathematical knowledge by assembling isolated, relatively unconnected mathematical statements, rules, and procedures. Thus, a Collector's calculus conceptualizations could be said to be a "collection" of statements, rules, and procedures. The external nature of Collectors' sources of conviction appeared related to the fact they approached their calculus learning as recall or rote memorization of statements, rules and procedures. Their sources of conviction were rooted in the validation made by others when these other individuals judged as correct the Collectors' statements and the Collectors' procedures. Although Collectors might validly apply calculus knowledge, they did not claim to know personally whether particular pieces of mathematics were valid or correct. Instead, they relied on others to determine validity or correctness. They perceived these other individuals as people who understand and find meaning in calculus.

Distinctive features of the interviews with Collectors are summarized in the following statements:

a. Collectors often unsuccessfully completed interview problems. They frequently made errors, displayed misconceptions, were unable to remember particular rules or procedures, or were unable to explain concepts.

b. Collectors explicitly stated that they approached their calculus learning by memorization of what they believed would be needed to pass an exam, while also stating they did not feel they personally understood calculus in terms of why one uses particular procedures or how procedures function in reaching a solution.

c. Collectors frequently made justifications by referring to statements, rules, and procedures they said they knew of because they had been given the information by a teacher or textbook. They claimed they did not understand them for themselves.

d. Collectors displayed beliefs that mathematics is a collection of definite, correct formulas, rules, and procedures.

e. Collectors often spoke of calculus as being separated from their reality and ways of understanding.

f. Most of the Collectors (all but two) explicitly expressed a lack of confidence in their abilities to personally understand calculus.

g. Collectors perceived calculus as different from previously studied mathematics. The above features of Collectors were clearly not independent of each other. For example, their beliefs that mathematics is a "black and white," "right or wrong" discipline were integrated with their beliefs that calculus is separate from their ways of coming to understand something. These points can be seen in the following interview extracts in which the students comment on their experiences learning calculus:

Doug: . . . something like English I can just do it. Political science. . . . There I can actually use, like I can just do it with my own mind. I can give my own interpretations of something. But in math it's either right or wrong.

Ned: . . . like with me I have to look through someone else's eyes. Like a foreign kind of viewpoint. That's very hard for me to do.

The fact that Collectors did not see as valid their personal ways of interpreting or expressing mathematics relates to beliefs that calculus was separate from their ways of understanding This devaluing of one's own mathematical interpretations was typical of Collectors and is particularly clear in the following reflections by Cindy and Daniel:

Cindy: . . . because I'm not perfect at interpreting it [calculus] in the correct mathematical language. Well I might be able to write it down, but it probably wouldn't be right. I probably wouldn't do it the correct way. But I would, if I was to go back and read it I would understand what I meant. But it wouldn't be the right way so anybody else would understand it.

Daniel: And right now it is a matter of being able to produce it [the correct answer] on a test. And whether or not my interpretation is correct doesn't matter. Because my interpretation isn't going to be counted on the test. . . . Well in most anything else I could feel confident my views are um maybe not necessarily correct, but that they're feasible, or that I can show how my views and somebody else's views correlate or something. Like you know. In math I don't feel that I have got any basis to say that I'm right and I'm wrong. Because if they, they referring to math people, come up with all this stuff, or how do I say it? I'm just not confident that my way of viewing it, like I could so easily be wrong.

In devaluing their own ways of making sense of calculus, these students did not allow internal sources of conviction to play a prominent role in building their calculus conceptualizations. Rather, they perceived truth and validity decisions in mathematics to be pre-determined entities external to themselves as learners. They did not generally see it possible that these decisions might be influenced by one's own perceptions or interpretations. Instead, they must be remembered or memorized.

Collectors' sense of alienation from the ways they felt they were expected to understand calculus raises questions about their perceptions of the context of their learning. Their comments of their experiences in learning calculus may indicate they felt pressures to be acculturated as opposed to enculturated [2]. That is, they saw learning calculus as a process of replacing their ideas with someone else's "better" ideas (acculturation), rather than a process of being encouraged to understand the mathematics via their existing world views (enculturation). Most noteworthy of Collectors, in relation to the notion of acculturation, is how strongly they expressed frustration when speaking of calculus as separate from their realities and ways of understanding:

Cindy: I mean normally in other math, like in the story of my math I have been. I've wanted to take the time to do things and figure out different things on my own. But I just found in this class I just got so frustrated. . . . Maybe I can't understand it the way I want to understand it. Like maybe, you know, like maybe because so much of it is abstract things, that you just have to be satisfied with not understanding. I don't know.

Daniel: Well he's a math professor. And I'm a political science major, university student. And I think the reason he's in math and I'm not, like that in itself is the fact that we both understand things differently. . . . and the way I describe things is more along a different line, you know, than mathematical notation. So if I told him an example about apples or pencils, he'd just kind of go X [wrong]. You know. It's just, like to him it's not what he wants you to know. And I have a hard time grasping. That's probably why I'm failing. . . . Oh, it's definitely different. But it's not, it's not, it's not mathematical. I mean it's just not. . . . So that's not acceptable.

What is not clear from these students' words is whether beliefs that mathematics is pre-determined and separate from one's interpretations led to externally oriented sources of conviction, or if external sources of conviction promoted perceiving calculus as separate from a sense of personal understanding. Most of the Collectors had been relatively successful with mathematics prior to studying calculus (final grade 12 mark of over 70%), and were generally achieving well in their other post-secondary courses. The first of these facts indicates, for these students, previously successful strategies for learning mathematics were no longer working. From a constructivist perspective the situation could be interpreted as students approaching their calculus learning with what previously were viable cognitive and social mathematical learning schema. However, in calculus, in their interactions with the subject inside and outside the classroom, Collectors previous schemas appear inadequate. That is, new demands were put on previously viable social, cognitive and metacognitive strategies. Within the social context of their calculus learning, students stated these demands largely resulted from needing "a lot more time in class":

Cindy: He [Professor Beta] would teach us a few, do a couple of examples, and then put a couple of problems on the board. Have us sit there and try to figure them out. And then have us put input into that. And tell him what we had problems with when we were trying to figure it out and that kind of thing. Then before we go home we would realize what we were having problems with. But now he teaches it all in one lesson and we go home and we do the assignment, and we find out there's problems there that he didn't either emphasize, or that we didn't realize there's problems. . . . But then he's already taught it and he's on to something else.

Cindy's words highlight the social interaction and time constraints Collectors experienced. More cognitively oriented demands arose in relation to their sense of calculus being separate from their ways of understanding. They saw calculus as "different from a lot of other mathematics" (Cindy) and "a new type of math from what we've learned all our lives" (Betty). Thus, the external nature of Collectors' sources of conviction may relate to a combination of constraints they experienced in their social learning environment and their view of calculus as "different" mathematics that is separate from their ways of making sense of experience.

Technicians.

Students classified as Technicians displayed a mixture of internal and external sources of conviction. Their external sources of conviction were similar to Collectors' in that they were based on knowledge of calculus statements, rules, and procedures. However, Technicians differed from Collectors in their perceptions and use of these statements, rules, and procedures. They perceived calculus as a logical organization of statements, rules and procedures, and they employed this organization as a technique for thinking about and applying calculus concepts. Therefore, what most distinguished Technicians from Collectors was their display of "personal" knowledge of how calculus statements, rules, and procedures fit together into a logical whole. For them, this logical whole became a calculus "technology" in that it is a science or method for thinking about and applying calculus. Technicians could therefore be viewed as skilled users of calculus techniques, and the role of their sources of conviction was as a set of tools that a technician employs to apply calculus concepts.

From a complete examination of the interviews the following major points were generated as indicative of Technicians:

a. Technicians generally displayed more extensive calculus knowledge and skills than Collectors, and generally completed the interview problems successfully.

b. Technicians' calculus conceptualizations were more organized and logically structured than Collectors'.

c. Technicians' sources of conviction were based upon statements, rules and procedures they saw as organized into a coherent, structured set.

d. Technicians displayed a mixture of external and internal sources of conviction, speaking of both memorizing rules and working through calculus problems for themselves.

e. Technicians expressed confidence in their mastery of application of the rules, and logical procedures of calculus.

f. Technicians saw calculus as different from mathematics they had studied previously.

The points above reflect how Technicians had built their conceptualizations as a logical structure of appropriate application of calculus ideas and techniques. This sense of knowledge of how calculus is structured as an applicable technology was what most distinguished them the most from Collectors. Instead of employing a collection of relatively unconnected mathematical statements, rules and procedures, Technicians saw calculus ideas and procedures as (borrowing their terms) something one can "build from" and employ as a "process" or "method of thinking" to "work through" and solve problems:

Jennifer: I think calculus, if you get into a method of thinking it's just a process. It seems to be the same sort of process and you just get into that method of thinking and it's all very logical.

Understand something? To take that tiny basis of logic and be able to build on it. Like using that maybe as a cornerstone. But if you understand that, then you can understand things more. . . . Then you can continue onto a higher level. . . . By applying to another concept. How can I say it? Through practical application.

For the other Technicians, as well as for Jennifer, it must be noted that, alongside an internal component related to a personal sense of mastery of the rules and logical procedures of calculus, Technicians' sources of conviction had a distinctly external component residing in statements, rules and procedures they perceived to have been generated externally. Technicians' beliefs about mathematics could therefore be said to overlap with the Collector's in that Technicians saw the statements, rules, and procedures of calculus as external to themselves as knowers. Their beliefs differed however from Collectors' in that their views of calculus as a "method of thinking" (Jennifer), a "pattern" (Richard) or a logical "step by step" (Sally) problem solving process were not present in Collectors' interviews. Their beliefs also differed in that they saw knowledge of application of calculus statements, rules and procedures as accessible to their sense of understanding. For example, Sally said that:

Sally: Understanding is applying the ideas to get a right answer. . . . And you need to know the ideas in order to apply them. And know what ideas apply in what circumstances.

A final point of contrast between Collectors and Technicians is that, whereas Collectors spoke of understanding things "along a different line" than mathematical notation, Technicians may have achieved some degree of understanding calculus language. In fact, their personal sense of mastery of the technology of calculus indicates they saw knowledge of the logical use of calculus language as a source of conviction. By coming to know how to use calculus symbols and terminology, they organized their calculus experiences and structured their related conceptualizations. However, they did not always perceive these conceptualizations to be personally meaningful. For example, Richard, who was achieving at a level of about 85% in his calculus course, said:

Richard: Well I just mean that in calculus when I say I understand something it means I can do it. And I can get the right answer. Whereas if I'm usually talking about another subject, well I understand the theory, or I understand the principles behind it. I know what is happening and I could, if somebody asked me to explain it to them, I could explain it to them in terms they could understand. Whereas I couldn't do that at all. I could never explain calculus to somebody in terms that they could understand. Because I don't understand. I just know how to do it.

Significantly Richard's words indicate it is possible for students to adopt the language use conventions and logical processes of calculus without also developing a personal sense of understanding of calculus. That is, they learn the "talk" of calculus, but "it doesn't seem real." "It's just talk. You can prove it but that's just talk too" (Sally). From a social constructivist perspective this lack of a personal sense of understanding calculus indicates Technicians' successfully acculturated with the cognitive structures and strategies of calculus promoted by the mathematics community at large. That this process was more validly described as acculturation rather than enculturation is evidenced by the fact that Technicians viewed calculus similarly to Collectors. They saw calculus as "a totally different kind of math" (Sally). "There's like calculus mode and there's a math mode type thing. And calculus is different from math" (Jennifer). Finally, the different nature of Technicians' and Collectors' sources of conviction appears to relate primarily to Technicians' views of calculus as a logical, organized structure, along with their capability to master the application of the components and procedures of this structure.

Connectors.

Students, who from their sources of conviction were classified as Connectors, displayed sources of conviction that were generally internal in nature; they resided in ideas and techniques Connectors experienced as making sense. That is, Connectors viewed calculus knowledge as something of which they could gain personal understanding and use. Similarly to Technicians, Connectors displayed knowledge of calculus as a technology. They organized their calculus experiences to be able to logically and consistently apply calculus ideas and techniques. However, Connectors differed from Technicians in that Connectors expressed a stronger sense of personal understanding of their calculus conceptualizations. Their conceptualizations were displayed as a network of "connections" among various aspects of calculus and between calculus and themselves. They used their internal sources of conviction to construct calculus conceptualizations they understood personally. In this way the role of Connectors' sources of conviction was as validation to them that they made statements, performed procedures, or created problem responses that were valid, correct and meaningful to themselves as well as to other individuals. Thus, Connectors were able to apply and make sense of calculus knowledge.

Connectors display the following prominent features:

a. A higher level of competence with calculus concepts and skills than most of the other students–their problem responses were often more detailed than the other students', using more symbolic representations and more complete explanations of ideas or procedures.

b. Connectors explicitly stated they approached their calculus learning by trying to understand for themselves and trying to connect together ideas, statements, rules, and procedures. They perceived calculus as something learned through personal involvement with and subsequent flexible application of ideas.

c. Connectors' sources of conviction were internal in nature; they resided in a sense of personal comprehension as well as a sense of control of calculus ideas and applications.

d. Connectors' sources of conviction were both a guide and a confirmation to them. They stated and used calculus ideas and applications in ways meaningful to themselves as well as to others knowledgeable in calculus.

e. Connectors expressed confidence in understanding calculus ideas and their abilities to apply calculus to a new problem.

f. Like Collectors and Technicians, Connectors spoke of mathematics as being definite and set in its ways, but also spoke of it as a practical or human endeavour.

The prominent factor distinguishing Connectors from Collectors and Technicians is their sense of themselves and their own thought processes and interaction with material as sources of conviction by which to learn and use calculus. For example, Connectors' sense of the importance of personal comprehension of calculus can be seen in the following interview excerpts in which Neil and Tanya comment on their calculus learning:

> Neil: I definitely try to recreate things and think it through. . . . Seeing how everything is linked together. And not just this idea, and this idea over here. And if they are connected then one should know it. Even if it's a little more complex, I think the connections are important.

> Tanya: Because you can't learn from memorizing everything. Because you have to interpret it. You have to understand the theory behind a certain form. The theory behind a certain something, and then apply it to something else. . . . Cause you need to, you need to imagine it in your head. What goes on. You can't, you can't see infinity. You have to imagine infinity. You can't see infinitely, or infinitesimally small. You have to imagine it.

Connectors saw their own interpretations and thought processes as components of their calculus learning. Their calculus conceptualizations were thereby constructed as a network of personally meaningful, interconnected statements, rules and procedures. As sources of conviction, they used knowledge and thought processes that were conceived of as their own.

Perhaps a connection exists between characteristics of Collectors, Technicians, and Connectors and such characteristics as "intelligence" or educational background. However, all the students came from very similar educational backgrounds, had completed a similar range of high school courses, all had achieved a grade 12 average of over 70%, and all but one (a Connector) spoke English as a first language. Thus, it is likely that the classification of students as Collectors, Technicians, or Connectors is more than merely a way of distinguishing "very good" from "weaker" students. The findings of this study indicate that it is a combination of belief structures about mathematics and what it means to learn mathematics, along with particular cognitive or metacognitive schema, that enabled Connectors to outperform Collectors.

These cognitive and metacognitive frameworks might be more mathematically advanced than those of Collectors with respect to what are appropriate cognitive conventions and strategies for successful achievement in calculus. The fact that Connectors felt they themselves were an integral part of the development of these connections and strategies indicates their learning could validly be described as a process of enculturation. They felt comfortable in adopting the thinking practices of calculus without necessarily abandoning their own beliefs and practices. That this description is in contrast to that of the "acculturation" of Collectors is of significance in relation to the implications of this study for future research and classroom practices.

Students' Language Use.

The results of the initial analysis of the number of occurrences of students' use of mathematical symbols, technical, or everyday language are reported in Table 1. The figures in Table 1 show both similarities and differences among the groups of students at the three post-secondary institutions. They also indicate the nature of students' language use. As mentioned, the role of students' language use was determined from more extensive examination of what students said or wrote and what this language reflected of their calculus conceptualizations. In the following sections, values from Table 1 are used as initial bases for discussion. The discussions are then expanded to include findings from examination of what students said or wrote in the interviews and what role these responses played in their interpretations of calculus problems. A more complete account of students' language use can be found in Frid [11].

Symbols. At all three institutions students made less use of symbols to represent, explain or justify ideas in comparison to their use of technical mathematical terminology or everyday language. Considering the highly symbolic nature of calculus this fact might be surprising. Furthermore, since about 75% of responses to skill problems were correct or within a couple of steps of being completed and correct, it can be concluded that although students were able to perform standard symbolic operations, they did not generally use symbols to convey ideas. That is, they were able to accurately manipulate symbols according to already learned rules, but they did not often use symbols to describe or explain a mathematical concept. In fact, even when asked to do so, students generally could not use symbolic representations in explanations. For example, many students could easily and accurately use derivative rules to manipulate symbols to find a derivative. However, few of these same students could use symbols to explain what the derivative is or to justify why a function is non-differentiable at a particular point. Ten of the 17 interview students used symbolic representations in one third or less of their problem responses, although symbolic representations were possible in all the problems. Many of the students' responses were constructed from knowledge of symbolic procedures and were accompanied by students verbalizing the steps they were going through. For example, Sally explained what she was doing in determining the derivative of $F(t) = (2t^2 + 3t - 2)^{10}(3t^{1/4} - 9)^7$ by verbalizing the steps she was using in the product and chain rules:

Sally: And then 3t. And then you use the multiplication rule for derivatives. Which is the derivative of the first one times the second term. Plus the first term times the derivative of the second term. So then we took the derivative of the first term. And then there's the second term. And then plus, and then this normal term. Times the derivative of the first term. So we first take the outer function. Times, and then the inner function.

TABLE 1. Language Type Totals and Institution
Averages for Students' Language Use

Instructional Approach Student	Symbols (S)	Technical Language (TL)	Everyday Language (EL)	EL/TL
Technique-oriented				
Annabel	7	23	14	.61
Ellen	2	8	24	3.0
Jennifer	2	22	30	1.4
Ned	3	24	37	1.5
Richard	5	29	22	.76
Averages	3.8	21.2	25.4	
Concepts-first				
Cindy	4	26	25	.96
Daniel	3	24	31	1.3
Doug	3	23	36	1.6
Leanne	4	30	27	.83
Sally	5	30	32	1.1
Tim	7	30	16	.53
Averages	4.3	27.2	27.2	
Infinitesimal				
Betty	8	16	43	2.7
Gordon	1	18	29	1.6
Mike	7	23	41	1.8
*Nadine	6	18	41	2.3
Neil	5	32	35	1.1
Tanya	7	42	45	1.1
Averages	5.7	24.8	39	

*The figures for Nadine are incomplete because there was insufficient time in her interview for one of the problems. These figures are therefore likely to be lower than they would have been if she had completed that problem.

Sally's response was typical of many of the students' problem responses. Whether they used technical terminology such as "first term times the derivative of the second term" (she presumably means "factor" here) or everyday language such as "outer" or "inner" function, students' language use appeared to serve them as a guide or recipe for completing a fairly well defined procedure (in this case to determine a derivative). Following the directions of the recipe achieved an answer. In other words, students' language use suggested that verbalized symbolic procedures served the students as tools by which to construct problem responses. This outcome is not surprising and is perhaps inevitable if one considers the skill reproduction orientation of a number of the questions posed. However, in conjunction with the fact that many of the students spoke of symbols as not having meaning for them, it reveals that masterful use of symbolic conventions and procedures does not automatically engender in students a sense of having constructed meanings for symbols. The students' lack of sense of having constructed meanings for concepts that particular symbols or combinations of symbols represent is reflected in the following sample interview extracts that arose when students were asked to comment on the meaning and role of symbols in their learning:

Daniel: Because to me it looks like Greek on the board when he works through all that stuff. Well usually the notation doesn't make a whole lot of sense sometimes. . . . like derivative is equal to f at a minus, or f at b minus f at a, all over b minus a. To me, what does that represent? It's a little too ah, like to me it's easier if you just say that. Instead of writing it out with as and bs. Like why don't they just say what they mean, you know?

Richard: They're just symbols I move around according to a rule. They don't really mean anything. . . . It doesn't have a meaning. It seems like it's stupid notation. Why don't they have notation that says what it is?

The above comments, as well as the other students' comments, revealed many of the students saw symbols as separate from any personally understood calculus conceptualizations. Most students who expressed this view were Collectors, indicating there might be a relationship between a lack of use of symbolic representations and approaching learning as a Collector.

An indication of a possible relationship between symbolic language use and one's approach to calculus learning was also seen in the problem responses of the six students (third of the group) who used symbols most (Annabel, Tim, Betty, Mike, Nadine, and Tanya). They each used at least 6 symbolic representations in their interviews (see Table 1). Numerically, these numbers of symbolic representations might not seem extensive. However, the representations used by this group of students differed in role compared to many of the others in that the representations were incorporated as an integral part of an explanation, rather than as a tool to proceed with a calculation. Four students in this group who used symbols most were Connectors. The two exceptions to this classification were Betty (a Collector) and Nadine (a Technician). However, these two students, as well as Mike and Tanya, were taught calculus using an infinitesimal approach to instruction. What is noteworthy about these students is that they, unlike most of the other students, were able to give symbolic justifications or explanations of continuity and differentiability. Furthermore, the symbols they used and their corresponding verbal language were particular to an infinitesimal approach to instruction. For example, part of Mike's response to Problem 5 was as follows:

(Problem 5)

[5. For each function given below, determine if it is continuous or discontinuous. Give reasons for your answer.]

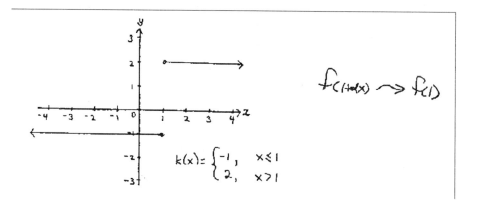

FIGURE 1.   Mike's Written Response to Problem 5.

Mike: So your dx could be positive or negative. Either side of it. But that would have to round off to the function at one. To be continuous. Now if you get a negative then it will. But if you get positive, all of a sudden you're going to . . .

I: If what's negative or positive?

Mike: Your dx. You know if you have negative dx you're going to be just to the left. And it's, well that rounds off to the correct thing. But just to the right of it, hey, I've got a two here, not a minus one.

This extract exemplifies how students can use infinitesimal language (words and symbols both) to explain the relationship between the behaviour of a graph and the corresponding notions of continuity or differentiability. In doing so, infinitesimal notation serves as a primary tool for construction of an explanation. It is a key tool in that interpretation of dx as an infinitesimal number provides one with something fairly concrete to work with. One can then easily visually locate on a graph what $\underline{dx}$ corresponds to and how the position of $\underline{dx}$ relates to the behaviour of the graph. In this way infinitesimal notation can be used to build and justify responses.

Traditional limit notation can be used to explain and justify continuity and differentiability in similar ways. However, of the eleven students who received calculus instruction using the limit concept, only Annabel and Tim (who were both Connectors) successfully justified continuity or differentiability by the use of limit notation. A third student, Jennifer (a Technician), displayed a sense she knew limits were needed to prove discontinuity, but she was unable to connect her ideas with the limit notation she wrote down. These findings suggest limit notation is not easily used by students as a tool for construction of calculus conceptualizations.

Technical and Everyday Language.

In relation to technical language use, students at all three institutions were similar in the extent to which they used technical language. The average use of technical terms was similar at all three institutions (see Table 1). These averages for technique-oriented, concepts-first and infinitesimal instruction were 21, 27, and 24, respectively. However, students who received infinitesimal instruction used, on average, 39 everyday language terms or phrases. The corresponding averages for students who received technique-oriented and concepts-first, instruction were 25 and 27, respectively. These findings indicate that although students who received infinitesimal instruction used technical language to about the same degree as students at the other two institutions, they used everyday language more. This fact distinguished them from the other students. What distinguished them more was the content of their technical and everyday language use and the role of this language in describing, explaining, or justifying calculus ideas. What follows is a summary of prominent features of students' technical and everyday language use. Also included are examples of students' responses from all three schools and a discussion of the nature and role of language in the interpretation of calculus problems.

Use of previous language knowledge. Students carried over to calculus previous conceptualizations associated with terminology. The powerful, influential role this previous language knowledge played in students' construction of calculus conceptualizations was particularly evident in students' explanations of the terms "round off," "limit" and "dis/continuous". For example, "round off" was associated with making numbers "easier to work with." Students' notions related to the round off process reflected ideas taught in elementary school, that rounding off makes numbers easier to work with. The students who were taught using limit notation and terminology ascribed meanings to limits by interpreting "limit" as an everyday language term, making reference to such things as barriers, swimming endurance, and borders. The explanations of "round off" given by students who received infinitesimal instruction were not congruent with the corresponding responses they gave to explain "limit." Thus, although the underlying notions of the concepts of "round off" and "limit" are synonymous, the language used to describe them appears to have guided students to construct conceptualizations that are not synonymous.

Integration of everyday and technical language. Everyday language was helped and hindered students' interpretations of calculus problems. Whether it was a help or a hindrance depended on the extent students integrated everyday language with technical language or symbols in ways congruent with the corresponding concepts. Students who received infinitesimal instruction, although they did not always make extensive use of symbolic representations, generally integrated everyday language more with symbols or technical language than did the other students. An example of this integration using language particular to infinitesimal calculus is given below.

(Problem 2)

[2. For each of the following sequences of numbers, decide whether the sequence rounds off to a particular number. If so, what is this number?

$$1, \quad \frac{1}{10}, \quad \frac{1}{100}, \quad \frac{1}{1000}, \quad \frac{1}{10000}, \quad \frac{1}{100000}, \quad \ldots$$

$$3.9, \quad 3.99, \quad 3.999, \quad 3.9999, \quad 3.99999, \quad 3.999999, \quad \ldots]$$

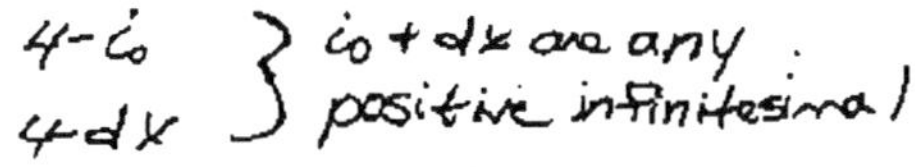

FIGURE 2. Tanya's Written Response to Problem 2.

Tanya: And it's going to get closer and closer to four. . . . Four minus an infinitesimal amount. Which isn't four, but it's as close to four as you get.

In this excerpt Tanya used the technical language term "infinitesimal" and the corresponding notion of an infinitesimal number to justify her previous explanation using the everyday language phrase "get closer and closer to four." For Tanya, technical terminology arising from an infinitesimal approach to calculus was connected to symbolic and everyday language use. In comparison, students who received technique-oriented or concepts-first instruction, although they often gave valid explanations of situations using everyday language, did not as frequently use technical terms or symbols for further, more detailed or precise justifications, even when asked to do so. In particular, unless specifically asked to do so, they did not make extensive use of language and ideas related to limits. Also on occasion, they used technical terms or symbols but were unable to explain the connections to everyday language explanations.

Use of visually oriented language. A prominent feature of interview responses from students at all three institutions was that students were often able to give valid explanations for problems when the problems were visually oriented by graphs. In particular, students used a range of technical and everyday language related to the visual appearance of a graph or the concepts of slope and tangent line. For example, the language they used included: "not smooth," "always decreasing," "tangent horizontal," "steepest," "turns sharply," "flat," "levels off," "stops" and "gets higher." This language was used to describe features of graphs and the role of this everyday and technical language was as a tool for construction of explanations. In relation to the use of visual, physically oriented language, one aspect of infinitesimal instruction displayed in students' problem responses and that appeared as important in their responses was the notion of magnifying a curve. At some point in their interviews all students who received infinitesimal instruction spoke of infinitely "magnifying" or "blowing up" the graph of a function. In an infinitesimal approach to calculus instruction magnification is a means by which a function can be examined "up close." The following interview extract exemplifies how this process works, and also displays related technical and everyday language and how they can play a role in a student's construction of problem responses.

(Problem 6):

[6. A friend of yours who recently completed high school mathematics is wondering what calculus is all about because he/she has heard you frequently use the word "derivative." What short explanations, sentences, or examples would you use to explain to your friend what the "derivative" is all about?]

Neil: If you were to magnify that function infinitely it would look like a straight line with the same point. And you could still have a rise and a run. Except the rise and the run would be infinitesimal as compared to a finite rise and run.

Non-infinitesimal language (limit language) related to the slope of a tangent line also served to orient students to descriptions of a curve. However, these descriptions, justifications and conclusions seldom made use of limit-related language or processes. In comparison, the notion of infinite magnification has limiting processes built into its use. This feature distinguishes it from traditional slope and tangent line notions in more than one way. First, it is a dynamic rather than static method for interpretation of graphs. Second, magnification makes the limit concept of "close to" accessible. That is, the visual mechanism of blowing up or infinitely magnifying a curve serves as a visual, physically accessible means to examine limiting notions. The traditional limit concept also has visual interpretations, but these were not used by students who received

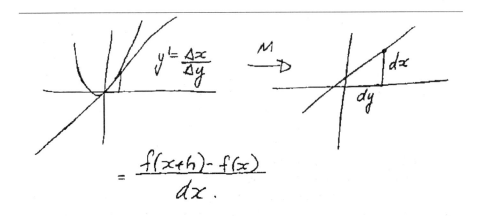

$$y' = \frac{\Delta x}{\Delta y}$$

$$= \frac{f(x+h) - f(x)}{dx}.$$

FIGURE 3.   Neil's Written Response to Problem 6.

technique-oriented or concepts-first instruction. In fact, the general absence of use of limit notation or terminology by students who received these approaches to instruction, unless it was specifically requested, indicates they did not integrate their limit conceptualizations into other calculus conceptualizations. For example, their responses included explanation of the derivative as the limit of the slopes of a sequence of secant lines, but the relationship of limits and derivatives was then not applied in other problem responses. Use of the notion of magnification was more regularly and comprehensively applied by students as a tool by which to construct calculus conceptualizations.

Conceptualizations built using infinitesimal language. Conceptualizations built using infinitesimal language displayed features different from conceptualizations built using traditional calculus language. More specifically, students who received infinitesimal instruction and who also used magnification and related terminology did not construct the same misconceptions present in problem responses of students who did not use infinitesimal terminology (including those students who received infinitesimal instruction but who did not use infinitesimal terminology). For example, students who did not use infinitesimal language tended to interpret the technical language term "continuous" in relation to everyday language phrases such as "no breaks," "no jumps," "existing," "being defined" and "not changing." Many of the students' notions associated with these everyday language phrases were valid interpretations of situations, although they were not necessarily valid mathematical interpretations. Therefore, the interpretations sometimes guided students to construct mathematically incorrect or inconsissstently used justifications. For example, interpreting continuity as "continually has a value" or "defined everywhere" led students to mathematically incorrect or inconsistent conceptualizations. In summary, students' problem responses revealed some important features of their technical and everyday language use. They built conceptualizations using infinitesimal language that were different from conceptualizations built using traditional calculus language. Whether speaking with traditional or infinitesimal language, students used terminology as foundation tools by which to conceptualize their constructions of descriptions and explanations, and in these constructions pre-calculus language knowledge was prominent.

CONCLUSIONS

The results of this study help clarify the nature of students' construction of calculus conceptualizations. Thus, this study takes steps towards helping educators tease apart reasons why many students do not understand calculus. It helps set the stage for facilitating calculus learning as a meaningful endeavour.

On a theoretical level, this study demonstrates the central notion of constructivism– that individuals construct conceptualizations that are viable models of their experiential world– is a practical, appropriate means by which to describe students' calculus learning. However, this study's findings show a constructivist perspective is not necessarily in accordance with the ways students themselves view calculus and their calculus learning. Students did not generally see mathematical knowledge as lying in the "shared rules conventions, understanding, and meanings of the individual members of society, and in their interactions" (Ernest, [10] p.82). For example, Collectors perceived calculus as separate from their own reality and own understandings. From a Collector's perspective there was no negotiation of calculus knowledge through interaction with other individuals. Instead, calculus learning through a Collector's eyes was a matter of replication of externally generated and independently existing statements, rules, and procedures. It was a process of acculturation to an alien culture (Bishop, [2]).

Since Connectors displayed a sense of personal understanding of calculus conceptualizations, their calculus knowledge might be interpreted as subjective knowledge of objective or public knowledge [10]. Alternatively, it could be interpreted as a process of enculturation [2]. However, most of the students were not Connectors. Thus, the search for effective ways to guide students to personal understandings of calculus must continue. Regardless of whether students apply calculus or study calculus beyond an introductory level, it is desirable that they pursue their calculus learning as a meaningful endeavour. Students who saw their calculus learning as personally understandable displayed more competence, confidence, and satisfaction in their abilities to do calculus.

Calculus instruction might be more successful for students if enabled students were more personally involved in the construction of their calculus conceptualizations. If instruction were designed from a social constructivist perspective to promote calculus learning as a process of subjective construction of publicly shared knowledge, then students more naturally might be guided to see building conceptualizations as a process using internal sources of conviction. In particular, this sharing should include a mutual sharing and negotiation between teachers and students of use of symbols and technical and everyday language phrases, along with personal interpretations of graphs, diagrams, statements and procedures.

The central role of language in mathematics learning is highlighted by this study's findings. Johnson's notions of "image schemata" are evidenced in this research study [18]. Image schemata are structures of meaning that arise from "perceptual interactions and bodily movements within our environment" (p.19). The image schemata evident in this study are those of continuity, slope, size, and magnification. Students from all three institutions made use of physical experiences such as continuity, slope, size, or magnification. In particular, magnification and its related notions of shape and closeness were used as image schemata by which to describe and explain key concepts. Gamma College students' use of infinite magnification in a variety of problem situations demonstrates that instruction emphasizing visual interpretations can impact students' conceptualizations; it can guide students to use bodily or kinesthetic experiences as sources of conviction. This use of bodily experiences in explaining calculus ideas indicates students are able to construct calculus meanings from physical experiences of the world. Instruction which emphasizes use of visual and physical calculus representations and their connections to symbolic reprsentations is therefore likely to enhance students' understandings of calculus.

Another aspect of language and learning discussed in the literature that this study supports is the role of natural, everyday language [15, 25]. Students' previous language experiences influenced their calculus conceptualizations. For example, students' pre-calculus knowledge of the terms "limit," "round off," "continuous," and "undefined" were evident in their calculus conceptualizations. This research study therefore further demonstrates that an individual's use and interpretation of everyday language is likely to figure significantly in that individual's mathematics learning. In addition, students' use of the technology of calculus can be interpreted as use of the calculus "register" [15, 25]. It is use of a register in that it involves the technical language of calculus, as well as characteristic modes of arguing.

The role symbol systems play in learning was also evidenced. Students' use of infinitesimal notation and related terminology was distinctly different from students' use of limit notation and related traditional terminology. Students who incorporated use of infinitesimal notation into their calculus problem responses used symbols as a "combined label and handle for identifying and manipulating concepts" (Skemp, [30]; p.62). They used infinitesimal language (words and symbols both) as tools by which to describe, explain and justify particular mathematical situations. In these instances their language use was a vital component of the construction of problem responses and related calculus conceptualizations. It oriented students to mathematically valid and useful descriptions, explanations and procedures.

Although definite conclusions cannot be made at this time as to the impact of different approaches to instruction on student learning, our findings point to potential avenues for further investigation of how instruction can influence students' language use and sources of conviction.

RECOMMENDATIONS FOR FURTHER RESEARCH

1. This study highlights the importance of students' beliefs about calculus and their related perceptions of their own calculus learning. Instruction aimed at improving student learning in calculus must address the influence of students' beliefs. Thus, future research needs to determine how to influence students' views of mathematics, perceptions of mathematics learning and approaches to learning mathematics.

2. Studies should be undertaken to investigate whether the three groups of students (Collectors, Technicians and Connectors) are present in other groups of students, both calculus students and other mathematics students. Such studies would be a step towards clarifying the combination of social and cognitive factors that give rise to the three groups.

3. This study indicated a relationship between students' perceptions of learning calculus and use of external vs internal sources of conviction. For the three groups, Collectors, Technicians, and Connectors, what is not clear at this point is the nature of the relationship between various characteristics they exhibit. In particular, since it is not clear if a causal relationship exists within the coexistence of certain features, more research is needed.

4. Research also needs to be conducted to determine if the nature of Collectors', Technicians', and Connectors' approaches to calculus learning might form a series of transitional learning phases. For example, it is not known if being a Technician might be a transitional phase between being a Collector and being a Connector. (A discussion of this possibility alongside a number of other issues related to the Collectors-Technicians-Connectors triadic structure can be found in Clarke, Frid and Barnett [4].

5. Mathematics teaching at all levels should investigate further the role of language conventions in creating new cognitive and social structures. That is, studies need to be conducted to determine how instructional language forms and related physical experiences can facilitate students' construction of personally meaningful and mathematically appropriate thought.

6. In particular, future research into students' calculus learning needs to examine in more detail students' use of visually oriented language and image schemata, their related use of symbolic representations, and their procedural use of symbols.

7. Finally, mathematics educators should re-examine both the goals and nature of calculus instruction. The alienation and frustration Collectors felt in relation to their experiences in learning calculus raise questions such as, is calculus, as it traditionally has been taught, largely a sifting mechanism for including certain kinds of thinkers and excluding others? Are there students who are inhibited in succeeding in calculus because they are required to acculturate? Are there alternative curricula which would facilitate calculus learning for all students as a process of enculturation?

Given the above points, it would be advantageous to continue research into calculus students' learning in particular, and mathematics students' language use and sources of conviction in general. Researchers and teachers would then be better guided in the development of mathematics instruction at all levels that facilitates proficient and meaningful mathematics learning.

Appendix A - A Sample of the Interview Problems

Problem 2.

(Technique-oriented and Concepts-first Instruction)

2. For each of the following sequences of numbers, decide whether the sequence has a limit. If so, what is this number?

$$1, \frac{1}{10}, \frac{1}{100}, \frac{1}{1000}, \frac{1}{10000}, \frac{1}{100000}, \cdots$$

$$3.9, 3.99, 3.999, 3.9999, 3.99999, 3.999999, \ldots$$

(Infinitesimal Instruction)

2. For each of the following sequences of numbers, decide whether the sequence rounds off to a particular number. If so, what is this number?

$$1, \frac{1}{10}, \frac{1}{100}, \frac{1}{1000}, \frac{1}{10000}, \frac{1}{100000}, \cdots$$

$$3.9, 3.99, 3.999, 3.9999, 3.99999, 3.999999, \ldots$$

Problem 3.

(Technique-oriented and concepts-first instruction)

(a) Evaluate the following:

$$\lim_{x \to \infty} \frac{x^4 + 4}{x^3 - x + 5}$$

(b) What does "limit" mean to you?

(Infinitesimal Instruction)

(a) Round off the following:

$$\frac{M^4 + 4}{M^3 - M + 5}$$

(b) What does "round off" mean to you?

Problem 4.

What can you say about the function $y = \dfrac{x^2 - 5x + 6}{x - 2}$ at $x = 2$?

Problem 6.

A friend of yours who recently completed high school mathematics is wondering what calculus is all about because he/she has heard you frequently use the word "derivative." What short explanations, sentences, or examples would you use to explain to your friend what the "derivative" is all about?

Problem 7.

Find the derivative of each of the following:

$$y = \frac{x^3 + \frac{1}{x}}{\sqrt{x} + 3x^2 + 7}$$

$$F(t) = (2t^2 + 3t - 2)^{10}(3t^{1/4} - 9)^7$$

Problem 8.

(Technique-oriented and Concepts-first Instruction)

What interpretations do you have for the expression below?

$$\lim_{h \to 0} \frac{f(x + h) - f(x)}{h}$$

(Infinitesimal Instruction)

What interpretations do you have for the expression below?

$$\frac{dy}{dx} = \frac{F(x + dx) - F(x)}{dx}$$

SANDRA FRID

Problem 9.

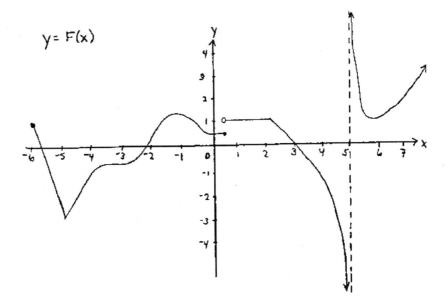

Problem 11. The number of elk in a national park at the beginning of each year is represented by the function $y = E(t)$ as shown on the graph below. The number of wolves is represented by the function $y = W(t)$, also graphed below.

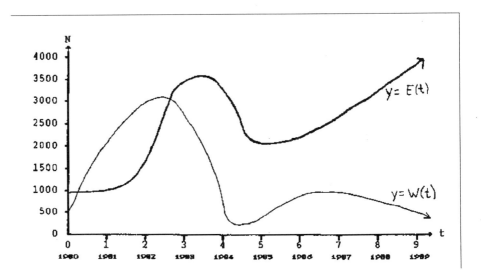

(a) At what exact point in time was the number of elk increasing most rapidly?

(b) During what time period was the rate of change of the number of elk decreasing?

(c)If you are told that for $0 < t < 4$ (ie. from 1980 to 1984) the equation for $y = g(t)$ is $W(t) = -100t3 + 1600t + 500$ (t measured in years), how would you determine all critical points of W?

(d) How would you use the critical points found in part (c) to determine the local and global extrema of W?

(e) At what point or points in time is the number of wolves not changing?

REFERENCES

1. Belenky, M., Clinchy, B., Goldberger. N. & Tarule, J., *Women's ways of knowing: The development of self, voice, and mind*, Basic Books, Inc., U.S.A., 1986.

2. Bishop, A., *Mathematical enculturation: A perspective on mathematics education.*, Kluwer Academic Publishers, Dortrecht, Netherlands, 1988.

3. Cipra, B., *Calculus: crisis looms in mathematics future*, Science **239** (1988), 1491–1492.

4. Clarke, D.J., Frid, S. & Barnett, C, *Triadic systems in education: categorical, cultural, or coincidence? Paper presented at the Annual Conference of the Mathematics Education Research Group of Australia, Brisbane, July 1993. (1993)..*

5. Croll, P., *Systemic classroom observation*, The Falmer Press, London, 1986.

6. Davis, R. & Vinner, S., *The notion of limit: some seemingly unavoidable misconception stages*, Journal of Mathematical Behavior **5(3)** (1986), 281–303.

7. Douglas, R. (Ed.), *Toward a lean and lively calculus: Notes on teaching Calculus-Report of the Methods Workshop* (1986), The Mathematical Association of America, xv-xxi.

8. Edge, O.P. & Friedberg, S.H., *Factors affecting achievement in the first course in calculus*, Journal of Experimental Education **52(3)** (1984), 136–140.

9. Ernest, P., *The constructivist perspective*, In Ernest, P. (Ed.). Mathematics Teaching: the State of the Art, The Falmer Press, London, 1989, pp. 151–152.

10. Ernest, P., *The philosophy of mathematics education* (1991), The Falmer Press, London.

11. Frid, S., *Undergraduate calculus Ssudents' language Use and sources of conviction. Unpublished Doctoral Dissertation* (1992), Department of Secondary Education,, University of Alberta, Canada..

12. Frid, S., *Undergraduate calculus students' sources of conviction. Paper presented at the Annual Meeting of the American Educational Research Association, Atlanta, Georgia, April 1993.* (1993).

13. Ginsburg, H., *The clinical interview in psychological research in mathematical thinking: Aims, rationales, techniques*, For the Learning of Mathematics **1(3)**, 4–11.

14. Goetz, J.P. & LeCompte, M.D., *Ethnography and qualitative design in educational research*, Academic Press, Inc., New York, 1981.

15. Haliday, M.A.K., *Sociolinguistic aspects of mathematics education. In Language as social semiotic.*, University Park Press, Baltimore, 1978.

16. Heid, K., *Resequencing skills and concepts in applied calculus using the computer as a tool*, Journal for Research in Mathematics Education **19(1)** (1988), 3–25.

17. Hirsch, C.R., Kapoor, S.F. & Laing, R.A., *Homework assignments, mathematical ability, and achievement in calculus*, Mathematics and Computer Education **17(1)** (1983), 51–57.

18. Johnson, M., *The body in the mind* (1987), University of Chicago Press, Chicago.

19. Kilpatrick J., *What constructivism might be in mathematics education In*, Bergeron, J.C., Herscovics, N. & Kieran, C. (Eds.)., vol. 1, Proceedings of the 11th International Conference of the Psychology of Mathematics Education, Montreal, 1987, pp. 3–27.

20. National Research Council, *Moving beyond myths, revitalizing undergraduate mathematics*, National Academy Press, Washington, D.C., 1991.

21. Orton, A., *Students' understanding of integration*, Educational Studies in Mathematics **14(1)** (1983), 1–18.

22. Orton, A., *Students' understanding of integration*, Educational Studies in Mathematics **14(3)** (1983), 235–250.

23. Peterson, I., *The troubled state of calculus*, Science News, **129(14)** (1986), 220–221.

24. Peterson, I., *Calculus reform: Catching the wave?*, Science News **132(20)** (1987), 317.

25. Pimm, D., *Speaking mathematically: Communication in mathematics classrooms*, Routledge, London, 1987.

26. Robinson, A., *Non-Standard Analysis*, North-Holland, New York, 1966.

27. Schuell, T., *Knowledge representation, cognitive structure, and school learning: A historical perspective.*, West, L. & Pines, A. (Eds.), Cognitive Structure and Conceptual Change,, Academic Press, Orlando, FL, 1985, pp. 117-129.

28. Selden, J., Mason, A., & Selden, A., *Can average calculus students solve non- routine problems?*, Journal of Mathematical Behavior **8(1)** (1989), 45–50.

29. Sierpinska, A., *Humanities students and epistemological obstacles related to limits*, Educational Studies in Mathematics **18** (1987), 371–397.

30. Skemp, R., *The psychology of learning mathematics. (Rev. American ed.)*, Erlbaum, Hillsdale, NJ, 1987.

31. Stewart, J., *Single Variable Calculus*, Brooks/Cole Publishing Company, California, 1987.

32. Tall, D., *Concept images, generic organizers, computers, and curriculum change*, For the Learning of Mathematics **9(3)** (1989), 37–42.

33. Tall, D., *Inconsistencies in the learning of calculus and analysis*, Focus on Learning Problems in Mathematics **12** (1990), 49–63.

34. Tall, D. & Vinner, S., *Concept image and concept definition in mathematics with particular reference to limits and continuity*, Educational Studies in Mathematics, **12(2)** (1981), 151–169.

35. von Glasersfeld, E., *An introduction to radical constructivism. In Watzlawick, P. (Ed). The Invented Reality*, Norton, New York, 1984, pp. 17–40.

36. von Glasersfeld, E., *Learning as a constructivist activity.*, Janvier, C. (Ed). Problems of Representation in the Teaching and Learning of Mathematics, Lawrence Erlbaum Associates, New Jersey, 1987, pp. 3–17.

37. Williams, S., *Models of limit held by college calculus students*, Journal for Research in Mathematics Education **22(3)** (1991), 219-236.

CURTIN UNIVERSITY OF TECHNOLOGY, PERTH, WESTERN AUSTRALIA

CBMS Issues in Mathematics Education
Volume 4, 1994

A Comparison of the Problem
Solving Performance of Students in
Lab Based and Traditional Calculus

JACK BOOKMAN & CHARLES P. FRIEDMAN

The Sloan Conference/Workshop on Calculus Instruction, January 2-6, 1986, better known as the Tulane conference, can be thought of as the "official" beginning of the movement to reform the teaching of calculus . Since that conference and the publication of *Toward a Lean and Lively Calculus,* the National Science Foundation has made a major effort to address the widespread dissatisfaction, documented in that volume and elsewhere, with the way calculus is traditionally taught [**3, 4**]. Duke University is the site of one of NSF's major, multi-year calculus reform projects. The program, called Project CALC [**1, 2, 5**] is radically different from the traditional college calculus course. The traditional course, taught to over 500,000 students each year, often emphasizes memorization and acquisition of pencil and paper computational skills, much of which can now be done by calculators and computers. Project CALC aims to give students a deeper understanding of the concepts of calculus by using interactive computer laboratories, small team investigation of real world calculus problems, cooperative learning, open book tests and extensive student writing [**5**].

Project CALC is one of several major calculus reform projects being implemented at colleges and universities across the United States. There is much interest in and debate about the issues and questions raised by these projects, yet there is little empirical evidence concerning the degree of their success or failure. This study provides data addressing one of these questions: To what extent has Project CALC met its goal of improving students' ability to solve problems?

This study arose from a view held by many of the students and by some of the teachers that students in the traditionally taught course were learning more. Because there was less emphasis on computational skills (what many

---

The research was funded by National Science Foundation grant# DUE-8953961.

undergraduates call "math"), many students in the Project CALC sections did not think they were learning. The purpose of this study was to see if the course was succeeding at its goals and, in particular, meeting the goal mentioned above. The data shows that, at the end of two semesters of calculus, students in the Project CALC sections performed significantly better on a set of five problems than students in the traditionally taught sections. These problems were designed to take between five and ten minutes to do and could be done with pencil, paper and a scientific calculator. The problems, discussed in more detail below, required students to formulate mathematical interpretations of verbal problems. The students were also asked to solve and interpret the results of some verbal problems that required calculus in their solution. Students in both courses, Project CALC and the traditional course, had been taught all the necessary content necessary to solve the problems, though the Project CALC course put greater emphasis on real world problem solving.

The purpose of this study was to measure the educational potency of the experimental course. In particular, we wanted to measure the impact of the course on the ability of students to solve non-routine word problems. This was an important goal of the course. This is a goal that would be considered by many to be a goal of the traditional course as well. In fact, the NCTM's [7] evaluation standards state:

The assessment of student's ability to use mathematics in solving problems should provide evidence that they can-
- *formulate problems;*
- *apply a variety of strategies to solve problems;*
- *solve problems;*
- *verify and interpret results;*
- *generalize results.*

These goals have been widely accepted by the mathematics education community and one would expect that they be a focus of any mathematics course, including new and traditional calculus courses. In short, we wanted to create a test whose items reflected these goals—items aligned with the goals of Project CALC yet be such that any well prepared calculus student should be able to solve them.

Background

Project CALC's goals, in broad and general terms, are that students should be able to:
- use mathematics to structure their understanding of and investigate questions in the world around them;
- use calculus to formulate problems, to solve problems, and to communicate their solutions of problems to others;
- use technology as an integral part of this process of formulation, solution, and communication, and

- work and learn cooperatively.

More specific goals for each of the three semester courses have also been developed [6]. The purpose of this study was to address the degree to which Project CALC succeeded at improving students ability to meet the second goal listed above and to meet the following more specific goals-that students should:

- understand the concept of rate of change and be able to formulate problems involving rates of change as initial value problems;
- solve initial value problems both numerically and formally and be able to explain and use the solutions;
- evaluate definite integrals numerically and formally, by computer and by hand.

Each week, the Project CALC classes meet for three 50 minute sessions and one two-hour laboratory session. The class time is divided among lectures, discussion and working in groups; these time allocations vary from instructor to instructor. In the lab, students work in pairs to explore real-world problems with real data, conjecture and test their conjectures, discuss their work with each other, and write up their results and conclusions on a technical word processor. This laboratory experience is a central component of the course. It shapes the contents and the approach of the text and the format of many of the classroom activities.

Project CALC is now in its fourth year at Duke University. During the first year of the project implementation, 1989-90, the two primary developers of the Project CALC each taught this experimental course to one section (approximately 20 students each). During the 1990-91 and 1991-92 academic years, approximately one third of the freshman calculus sections, were taught using the Project CALC methods and materials. Beginning in the 1992-93 year, all students beginning with Calculus I will take the new calculus course. In addition, during 1990-91, Project CALC materials were used at seven other colleges serving as test sites. During the academic year 1992-93, 25 other schools (2- and 4-year colleges, universities, and one high school) were using the materials.

The students with whom we compared the Project CALC students took a "traditional" calculus course that met three hours per week (as opposed to the five hours per week for Project CALC). There was a greater emphasis on skill acquisition in the traditional course than in the Project CALC classes. The traditional course covers more material and at a greater level of technical difficulty than does the Project CALC course.

An important part of the course development was a plan for formative and summative evaluation. The overall evaluation plan for Project CALC had four main components. This paper will focus on the first component of the evaluation. The four parts were:

1. A problem solving test given to students currently enrolled in the second semester of Calculus, both Project CALC (PC) and traditional (TR). This test was given in the spring of the second and third year of the development of the new course.

2. A study of sophomores and juniors comparing those who had taken PC and TR on five long term outcomes of calculus instruction: writing, attitudes, skills, concepts, and problem solving. This test was given in the third year of the project.

3. A comparison of the activities and attitudes of TR and PC students in their classes, and what they and the faculty liked and disliked about their course. This was done throughout the development of the project but most intensively during the second year as part of the formative evaluation process.

4. A follow-up study focused on the question "Do Project CALC students do better in and/or take more courses that require calculus?" In the fourth year of the project (1992-93), the math and math-related grades of the PC subjects with those of TR students who took Calculus I and II in academic year 90-91 was examined. In addition, interviews were conducted of pairs of students — one PC, one TR, matched by major and SAT scores — to see how they perceive their experience in mathematics at Duke and to inquire about their feelings about how well they were prepared for future mathematics related courses.

Method

In spring 1991, a short five question test of problem solving was developed and administered to all the Calculus II students of the project evaluator, 23 of whom were PC students and 42 of whom were TR students. Experiment 1 described in this paper refers to this test. In April of the following year, a revised version of the problem solving test was given to all Calculus II students, 65 in PC and 205 in TR.

The test items were selected with contexts from various fields-biology, chemistry, economics as well as mathematics. This was done to address the goal of using mathematics to investigate real world questions. As many fields as possible were selected so that the test would not be biased in favor of a group of students with any one academic background. The problems were also selected, particularly in the second version of the test, to meet two criteria: (1) that they be novel to both groups of students and (2) that they reflect topics that were taught in both courses.

We hypothesized that PC students would do better than TR students on this test. The PC course was designed to provide students with many more opportunities to solve problems than they would have in the TR course. A significant difference in favor of the PC students would be a necessary but not sufficient condition for considering whether the course is successful.

Students were originally randomly assigned to Calculus I sections, but were allowed, in accordance with university policy, to switch sections during the first two weeks of classes. In general, the subjects were very good students. Most of them had combined SAT scores above 1200. Most also had some form of calculus in high school, although they did not receive advanced placement credit. The gender of each of the subjects in Experiment 1 was known to the authors, but this information was not available in Experiment 2. Therefore the effect of gender was analyzed for Experiment 1 only.

The study was conducted twice in each of two consecutive years. Experiment 1 used an initial version and was given in the spring of 91 to the PC and TR students of one instructor. The following year the test was revised to response to analysis of the first trial. Experiment 2 is essentially a replication of Experiment 1 using the revised test and a larger number of subjects.

Experiment 1

In 1990-91, one of the authors (JB) taught one section of Project CALC ($n = 23$) and two sections of traditional calculus ($n = 42$). On April 16, 1991, students in all three sections completed an exam consisting of the following questions.

PROBLEM SOLVING TEST FOR EXPERIMENT 1

1. Suppose a cell is suspended in a solution containing a solute of constant concentration C_s. Suppose further that the cell has constant volume V and the area of its permeable membrane is the constant A. By Fick's law (Adolf Fick was a nineteenth century physiologist), the rate of change of the cell's mass m is directly proportional to the area A and to the difference $C_s - C(t)$, where $C(t)$ is the concentration of the solute inside the cell at any time t. Write a differential equation that can be used to find $C(t)$ if the mass $m = VC(t)$. YOU DO NOT NEED TO SOLVE THE DIFFERENTIAL EQUATION.

2. Five grams of a toxic substance are ingested by a person. The person's kidneys will eliminate the substance from the body at a rate proportional to the amount present. After one hour, one gram is eliminated. How long will it take before 90% of the substance is gone from the body?

3. Suppose $\dfrac{dy}{dt} = \ln(y^2 + 1)$ and $y = 0$ when $t = 0$. Describe how you might find (either approximately or exactly) the value of y when $t = 3$. Explain your answer clearly and be as specific as possible.

4. Consider a chemical reaction in which two substances combine to form a third substance. If this results from collision of the molecules of the two substances, then the Law of Mass Action states that the rate of formation of the new compound is proportional to the number of collisions per unit time, and that this number is jointly proportional to the amounts of the original substances not yet transformed. (*Note: A is jointly proportional to B and C if A is proportional to the product of B and C.*)

Suppose that y grams of the new compound is made up of $\frac{1}{3}y$ grams of the first reactant and $\frac{2}{3}y$ grams of the second. Suppose further that there were initially 12 grams of the first substance and 30 grams of the second. The Law of Mass Action leads to a differential equation whose solution is:

$$y = \frac{360 - 360e^{-kt}}{10 - 8e^{-kt}}$$

What can you say about the limiting value of y as t becomes large? Does your answer make sense in terms of the physical limitations of the reaction? Explain.

5. Suppose you are told what the net worth of a pension fund was on the first day of each month and each year from January 1, 1970 to January 1, 1990. You are also provided with a rough sketch of the graph of the data points (Figure 1):

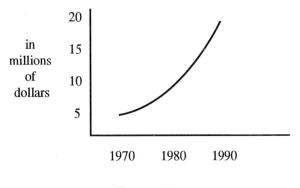

FIGURE 1

How would you go about predicting the net worth of the fund on January 1, 1995? Obviously we haven't given you enough information here to do this, but describe what steps you might take to answer the question.

END OF PROBLEM SOLVING TEST FOR EXPERIMENT 1

The students in both groups took the test at the same time in the same location. The test was given in the evening when students would have no conflicting classes. The students were given two hours, which was much more than was needed. The test was open book, open notes (Of course, the different groups had different books). Each question was graded on a five point 0–4 scale where: 0 meant no significant progress; 1 meant the student had made good start but no progress afterwards; 2 meant some good work had been done but either several minor details or a concept were missing; 3 implied a good job had been done, except for minor details; and 4 meant the work was completely correct. Each test was graded by two independent graders who were blind to group member-ship. All exams were mixed and randomly ordered prior to grading. Interrater agreement was high ($\alpha = .94$).

The test consisted of five questions novel to both groups. The test questions were based on five stated goals and objectives of Project CALC stressing appli-cation and understanding of concepts over technique. Question 1 asked students to set up but not solve a differential equation involving an example from cell biology. This question addressed the goal of whether students were able to for-mulate problems involving rates of change as differential equations. Question 2 asked students to solve an exponential decay problem (typical also of problems found in traditional calculus) involving a setting from biology. This question addressed the objective of formally solving initial value problems. Question 3 asked students to describe how they might find (either approximately or exactly) a solution to an initial value problem, given in a purely mathematical setting. Since question 3 cannot be answered by antidifferentiation, this question ad-dressed the objective of solving initial value problems numerically. Question 4 asked students to interpret a (given) solution to a differential equation using a setting from chemistry. This question addressed the objective of being able to explain and use the solutions of initial value problems. In question 5, students were given a rough sketch of the graph of some data points (from an example in finance) and were asked to describe the steps they might take to predict values of a particular data point not given. The purpose of this question was to test the students' understanding of functions.

Experiment 2

One year later, in April 1992, we repeated Experiment 1 with a revised test that involved as many calculus classes as possible. The test was given to three sections of PC and 11 sections of TR. Students who had switched from PC sections in the fall to TR sections in the spring were excluded from the study. No fall TR students had been allowed to register for PC calculus in the spring. Sixty-five PC students and 205 TR students completed the test. The average SAT scores of students in all the PC and TR calculus sections for Spring 92 are given in Table 1.

TABLE 1. SAT Scores of Subjects in Experiment 2

	N	SAT Verbal		SAT Math	
		Mean	SD	Mean	SD
PC	66	599	82	665	75
TR	269	604	67	676	58

As measured by SAT scores, the two groups of students were of equivalent general academic ability.

Unlike Experiment 1 where all subjects had the same teacher, subjects in Experiment 2 came from classes with different teachers. Students were told that this test would count a small but significant amount of their grade. The students were also told: (1) that the purpose of the test was to evaluate their skill at problem solving, and (2) that it was not necessary to study for it. The test was an open-book exercise consisting of the five problems shown in Figure 2. The students took the test during one fifty minute class period which was enough time to complete it. The test was graded on the same five point (0–4) scale as in Experiment 1, and as before, all exams were mixed and randomly ordered prior to grading. Because of the cost of using two graders and because of the high interrater reliability found in Experiment 1, one grader was used in Experiment 2.

PROBLEM SOLVING TEST FOR EXPERIMENT 2

1. Suppose $p(t)$ gives the rate at which telephone calls come into a regional switching center at time t hours where t goes from 0 to 24, starting at midnight. In this context concerning telephone calls, give an interpretation of

$$\int_9^{12} p(t) \, dt.$$

2. Suppose a cell is suspended in a solution containing a solute of constant concentration C_s. Suppose further that the cell has constant volume V and the area of its permeable membrane is the constant A. By Fick's law (Adolf Fick was a nineteenth century physiologist), the rate of change of the cell's mass m is directly proportional to the area A and to the difference $C_s - C(t)$, where $C(t)$ is the concentration of the solute inside the cell at any time t. If the mass $m = VC(t)$, write a differential equation that can be used to find $C(t)$ (i.e. write a differential equation of the form $\dfrac{dC}{dt} = $ ____).
YOU DO NOT NEED TO SOLVE THE DIFFERENTIAL EQUATION.

3. (a) Consider a chemical reaction in which two substances combine to form a third substance. If this results from collision of the molecules of the two substances, then the Law of Mass Action states that the rate of formation of the new compound is proportional to the number of collisions per unit time, and that this number is proportional to the products of the amounts of the original substances not yet transformed. Suppose that y grams of the new compound (call it C) is made up of $\frac{1}{3}y$ grams of the first reactant (call it A) and $\frac{2}{3}y$ grams of the second (call it B). Suppose further that there were initially 12 grams of A and 30 grams of B. If the reaction is allowed to progress for a very long time, about how many grams of A, of B, and of C will there be at the end of that time.

(b) Someone asserts that the Law of Mass Action leads to a differential equation whose solution is:

$$y = \frac{360 - 360e^{-kt}}{10 - 8e^{-kt}}.$$

Given your answer in part (a), does this assertion make sense? Why or why not? Explain briefly.

4. If you were asked to find $\int_0^4 5^{\sqrt{x}}\, dx$ exactly, you could not do it because you don't know an antiderivative of $5^{\sqrt{x}}$. However you should be able to estimate the answer. Is it:

(a) less than 0?   (b) 0 to 9.999?   (c) 10 to 99.999?
(d) 100 to 999.999?   (e) over 1,000?

Explain why you chose the answer you did.

5. A company is considering two ways to depreciate a piece of capital equipment that originally cost $14000 and is worth $10000 after <u>one</u> year:

Method I assumes it depreciates at a rate proportional to the difference between its value and its scrap value of $400.

Method II assumes it depreciates linearly, i.e. it depreciates at a constant number of dollars per year.

Find the value at the end of :

2 years using Method I: _____   2 years using Method II: _____
3 years using Method I: _____   3 years using Method II: _____

Which method produces "faster" depreciation? Explain.

END OF PROBLEM SOLVING TEST FOR EXPERIMENT 2

The goals of the questions in Experiment 2 were similar to those of Experiment 1, but several significant changes were made in the questions to address problems discovered in the first administration of the test. Question 3 of Experiment 1 was dropped because the wording of the problem was more familiar to the PC students than to the TR students. It was replaced with another numerical problem (see problem 4 in Figure 2) that was in a form that would be familiar to all students in Calculus II. Question 5, the least clear question of Experiment 1, was dropped and replaced with a very different problem (see question 1 in Figure 2). The context, but not the mathematical content, of question 2 of Experiment 1 was changed (see question 5 in Figure 2) to make the problem less isomorphic in form to textbook exponential growth problems. The other two questions remained essentially the same, though there were some changes in format and wording.

Results

Experiment 1. The internal consistency reliability (α) for the five problem exam was .54 reflecting the small number of items. Table 2 reports and compares mean scores for groups as computed by averaging the scores of the two graders. Out of a possible 20 points, PC students averaged 9.7 and TR students averaged 7.1. This yielded an effect size of .72 standard deviations. The difference of the means was significant at the .01 level. There were significant differences in the means of the scores of items 1 and 5 .

TABLE 2. Results of Experiment 1

	PC ($n = 23$)		TR ($n = 42$)			
	mean	sd	mean	sd	t-statistic	p-value
Total score:	9.	4.0	7.1	3.1	2.897	.005
Question 1:	2.1	1.3	1.3	1.3	2.296	.025
Question 2:	2.6	1.5	2.8	1.3	-0.560	.578
Question 3:	0.8	1.2	.45	0.6	1.619	.111
Question 4:	2.7	1.4	2.0	1.5	1.773	.081
Question 5:	1.5	.99	.52	0.5	5.282	.000

Gender did not seem to contribute to the difference in the performance of the two groups. Table 3 reports and compares the total scores for each group by gender and the results of the two way ANOVA. For both males and females, PC students performed about 2 points better than the TR, with effect sizes greater than .5. In each group, males scored about 2 points better than the females. The ANOVA shows that there was no interaction between treatment and gender and that there was a significant group effect for treatment even when gender was factored in.

TABLE 3. Results of Experiment 1 by Gender

males:	PC (n= 14)		TR (n= 22)			
	mean	sd	mean	sd	t-statistic	p-value
Total score:	10.4	4.2	8.0	2.1	2.253	.031
females:	PC ($n = 9$)		TR ($n = 20$)			
	mean	sd	mean	sd	t-statistic	p-value
Total score:	8.556	3.7	6.1	3.7	1.687	.103
Analysis of Variance						
Source	Sum of squares	DF	Mean square	F ratio	P	
group	48.589	1	48.589	4.310	.042	
gender	40.676	1	40.676	3.608	.062	
group × gender	0.034	1	0.034	0.003	.956	
error	687.743	61	11.274			

Experiment 2. The internal consistency reliability coefficient for the revised five problem exam was .52. Table 4 reports and compares mean scores for groups in the second experiment. Out of a possible 20 points, PC students averaged 10.8 and TR students averaged 8.4. This yielded an effect size of .55 standard deviations. The difference of the means was significant at the .01 level. There were significant differences in the means of the scores of items 2, 4 and 5 .

TABLE 4. Results of Experiment 2

	PC (n= 65)		TR (n= 205)			
	mean	sd	mean	sd	t-statistic	p-value
Total score:	10.8	3.8	8.4	4.3	3.932	.000
Question 1:	3.4	1.4	3.0	1.6	1.589	.113
Question 2:	2.1	1.4	1.2	1.2	3.740	.000
Question 3:	0.9	1.3	0.8	1.4	0.295	.768
Question 4:	2.6	1.9	1.8	1.8	3.023	.003
Question 5:	1.8	1.2	1.1	1.1	2.927	.004

Discussion

Experiment 1. On Question 1 PC students significantly outperformed the TR's. There was no specific content or lesson in the PC course that would explain why they would do better; the difference could be related to the difference in pedagogy and the general nature of the exercises assigned.

On Question 2, the TR students did slightly better than did the PC students. One explanation is that the TR students had solved many problems just like this one about two months before the test; the PC students had not seen problems of this type in about six months.

In Question 3, PC students would be expected to outperform the TR students since Euler's method was not discussed in the TR course and was discussed several times in PC. Most students in both groups tried to antidifferentiate either $\ln(y^2 + 1)$ or $\dfrac{1}{\ln(y^2 + 1)}$. It is not possible to find either of those exactly. In the event, both groups performed poorly. Perhaps, the problem should have been worded more clearly to cue the students that a numerical solution was needed.

In Question 4, though there was not a statistically significant difference between the two groups, the PC students outperformed the TR's. There was no specific content or lesson in the PC course that would explain why they would do better. Any difference would probably be related to the difference in pedagogy and the general nature of the exercises assigned.

In Question 5, the graph of the data suggested exponential growth, so many of the students assumed the function was exponential. Many fewer PC students made this unjustified assumption. The PC students scored significantly better than TR students.

Looking at this test as a whole, PC students outperformed the TR's. Because the test was skewed toward Project CALC's goals, this is not a surprising finding but it does indicate that Project CALC has made some progress towards meeting its goals. Improvements in the course and greater experience on the part of the instructors may widen the differences between the two groups. A longer test with a higher reliability would better estimate the magnitude of the difference.

Experiment 2. The results of the second study are consistent with those of Experiment 1. Though neither group did particularly well, the difference between the two groups was significant. PC students scored higher than TR students on each problem, with significant differences on three of the five questions and significant differences in the total score. While it is not surprising, given Project CALC's emphasis on differential equations, that PC subjects outperformed TRs on question 2, it is worth noting the significant differences on questions 4 and 5. Exponential growth (the subject of question 5) is covered in the second semester of the TR course and in the first semester of the PC course, yet despite the more recent exposure, TR subjects did significantly worse. Simpson's rule and the trapezoid rule which are two of the methods that could be used to solve question 4 are topics in the syllabus of the second semester TR course. On question 4, many of the TR subjects tried to come up with a symbolic rather than numerical solution to the problem. Question 3 proved to be very difficult for both groups of students; there were a lot of factors and ideas to keep track of, creating a cognitive load that, apparently, these students were not able or willing to overcome.

To put these findings in perspective, one of the faculty members who taught three sections of the TR group, asked her classes to write down, on the class day following the administration of the test, what they thought of the test: "to provide constructive criticism" of the test. A sample of the comments from these TR students is given below.

"I thought the problem solving test was easy once I figured out that all of the answers were just in thought (not in real numerical form). For people who were in the mind set that they needed numerical answers, I don't think the questions were asked specifically enough."

"I just think more of these kind of problems should be done in class."

"I think for the most part the problem solving was a fair assessment of the material learned in Calc I and II. However with no prior explanation of the nature of the test, it was quite difficult in a 50 minute period to recall all the techniques we have used. "

"I think the math department does a very poor job in preparing us for 'real world' problems. ... I don't think the concepts are adequately taught. I think we tend to rush over material in an attempt to meet 'set' deadlines." As a result, we only gain a cursory understanding of the information. I don't think the math department should switch over to all Project CALC, but I do think that calculus should meet 4 times a week in order to cover the material thoroughly."

"I liked the problem solving test. I normally like and do better on word problems. The questions were a little tricky but I felt there was nothing on the test that I hadn't seen before. I think the reason why we didn't do as well was because we weren't accustomed to the questions and the way they were asked."

"I believe that this test was designed to favor Project CALC students. The wording was hard to understand at times which completely misled me into setting up the problems incorrectly."

"Overall, I think I would have done better if we had taken this test immediately after learning the law of exponential growth, and integrals. The material would have been fresher in my mind − instead, I was thinking in terms of series."

"I don't think this test was appropriate to test regular calculus against project because it was a project calc test. If they were given a regular test, they would do as bad as we did. I have friends in project [calc] who can't integrate or derivate [sic] actual numbers."

Many students complained that the wording was difficult or ambiguous, as in the following comment: "word problem was confusing − there are too many letters and relations to separate." Other comments concerned the fact that the test was given on a hot Friday afternoon and that it didn't count much in their grade. Another student made what might be a very perceptive comment: "You could tell that these were science problems written by mathematicians. The sentences didn't make clear what was important and what was not."

Some of the underlying beliefs about mathematics held by some of the students can be inferred from these comments. Many of the TR subjects viewed the problems as "Project CALC problems", i.e. they equated word problems and problem solving with Project CALC and not their course, even though the TR course did include all the mathematical tasks necessary to solve the problems.

Conclusions

Looking at these tests as a whole, PC students outperformed the TR students. Because the test was skewed toward Project CALC's goals, this is not a surprising finding but it does indicate that Project CALC has helped students achieve at a higher level on the goals that are central to the vision of the course despite the doubts expressed by many of the students while they are taking the course. The data support the hypothesis that we can get measurable improvement on a pencil and paper test of problem solving by making it a major emphasis of the course.

These results were obtained during the second and third year of the development of the course when many of the instructors were teaching the course for the first time. Improvements in the course and greater experience on the part of the instructors may widen the differences between the two groups. A longer test might better estimate the magnitude of the difference.

We are not asserting here that the new course is a globally better course than the traditional one, rather that it is delivering some of what it promises. It is important to note that the data presented here are only one aspect of a much larger plan to evaluate the success of this curriculum reform. There are many other issues that need to be addressed. Some, such as student attitudes, skill acquisition, conceptual understanding and success in future math and science classes, are being addressed in the larger evaluation of this course; other issues, such as cost effectiveness, are not. It is also important to note that it takes time to develop and evaluate major curricular reforms. Even with the multicomponent evaluation we have conducted, we can only consider a sample of the complex social, psychological and intellectual endeavor of an educational experience.

The evaluation of Project CALC conducted over the past four years has pointed out areas of success as well as areas where the course needs to be improved. These conclusions are summarized below and will be reported more fully elsewhere.

Strengths of the project.

•As noted in the study reported in this paper, students in PC became better problem solvers in the sense that they were better able to formulate mathematical interpretations of verbal problems and solve and interpret the results of some verbal problems that required calculus in their solution.

•Although initially most students dislike the course, their attitudes gradually change. When surveyed one and two years after the course, PC students felt, significantly more so than the TR students, that they better understood how math was used and that they had been required to understand math rather than memorize formulas.

•In the classroom, PC students are much more actively engaged.

Weaknesses of the project.

•Initially student reactions to PC are quite negative. The course violates their deeply held beliefs about what mathematics is and asks them to give up or adapt their coping strategies for dealing with mathematics courses. This, of course, is not entirely a weakness.

•PC students do less well on computational skills involving symbolic manipulation. This has been the main emphasis of the TR course. Whereas, there is a danger in overemphasizing computational skills at the expense of other things, PC students should leave the course with better abilities in this area. It is the opinion of the faculty that this problem can be remedied by including more practice with routine calculations.

•The first large group of students of PC students (1990-1991 freshman) did slightly less well in future math and math related courses. Specifically, the students from that entering class who had a year of PC and at least two more math related courses by the middle of their junior year, when compared with students from that class who had a year of TR calculus and who also had at least two more math related courses by the middle of their junior year, did on the average about .2 grade points worse per course using the standard 4 point grade point system. An interpretation of this result is that PC students are being less well prepared for courses that are taught in a traditional fashion; this may have no bearing on whether they will be better scientists, engineers, or economists.

There is much to say about student attitudes concerning the calculus reform movement, more than can reported here. Briefly though, what seems to be the case is that initially student reactions to experimental courses such as the one described here are quite negative. The course violates their deeply held beliefs about what mathematics is and asks them to give up or adapt their coping strategies for dealing with mathematics courses. Often, the better students (like the subjects in this study) are more resistant to change precisely because those strategies have gotten them where they think they want to be. Beliefs such as: "Mathematical thinking consists of being able to learn, remember, and apply facts, rules, formulas and procedures" [8] and "One succeeds in school by performing the tasks, to the letter, as described by the teacher" [9] are at the root of the initial student resistance to reform of mathematics education and contribute to the belief discussed above that they are not "doing math." It was these beliefs that helped form the conditions that motivated this study.

We have found that it takes at least a semester (with great variance) for most students to begin to accept new ways of looking at mathematics. The change is gradual and difficult and when given an opportunity, students will easily lapse into the more familiar and comfortable routines. We have found that the results of this study have had a positive influence on changing attitudes about the course among both faculty and students.

Our next task is to explain what in the learning environment of Project CALC helped produce these results. This will require a deeper understanding of how students learn mathematics and this in turn will require the development of new theories of how students, and in particular how college students, learn mathematics. We can say that students in the Project CALC course get a lot more practice in reading and writing mathematics and doing mathematical problems in verbal contexts. The instruments in this study—the pencil and paper tests—were product or outcome measures. Methods of studying the mathematical development of students over time and instruments that addressed the mental processes of subjects would be helpful in gaining further understanding of the results found in this study.

REFERENCES

1. Gopen, G.D. and Smith, D.A., *What's an Assignment Like You Doing in a Course Like This.*, College Mathematics Journal **21** (1990), 2–19.
2. Smith, D.A. and Moore, L.C., *Project CALC.*, Tucker, T.W. (ed) Priming the Calculus Pump: Innovation and Resources, Mathematical Association of America, 1990.

3. Douglas, Ronald G. (ed), *Toward a Lean and Lively Calculus*, Mathematical Association of America, 1986.

4. Steen, Lynn (ed)., *Calculus for a New Century*, Mathematical Association of America, 1988.

5. Moore, L. and Smith, D., *Project CALC: Calculus as a Laboratory Course*, Tomek, I. (ed.) Computer Assisted Learning, Springer-Verlag, 1992.

6. Moore, L. and Smith, D., *Project CALC Instructor's Guide*, D.C. Heath and Co., 1992.

7. National Council of Teachers of Mathematics, *Curriculum and Evaluation Standards for School Mathematics*, The Council, Reston, VA., 1989.

8. Garofalo, Joe, *Beliefs and Their Influence on Mathematical Performance*, The Mathematics Teacher **82** (1989), 502–505.

9. Schoenfeld, Alan, *When Good Teaching Leads to Bad Results: The Disasters of "Well-Taught" Mathematics Courses.*, Educational Psychologist **23** (1988), 145–166.

DUKE UNIVERSITY

UNIVERSITY OF NORTH CAROLINA

CBMS Issues in Mathematics Education
Volume 4, 1994

An Efficacy Study of the Calculus Workshop Model

MARTIN VERN BONSANGUE

Achievement in mathematics has traditionally served as an international measure of the effectiveness and prowess of American academic institutions [11, 35, 36]. Calculus, the entry-level mathematics requirement for science-based majors, has particularly been associated with intellectual validation and financial promise [8, 19]. However, increasing failure rates in mathematics courses, particularly among underrepresented groups, have become a matter of national concern, both academically and politically [23]. During the academic year 1989–1990, for example, 47 bachelors degrees and 1 masters degree in mathematics were awarded to African-American, Latino, or Native American students throughout the entire California State University system [2]. Data from the National Center for Education Statistics indicates that increasing diversity among the population of undergraduate students majoring in mathematics- based disciplines has been accompanied by attrition rates that have remained disproportional across gender, ethnic, and age groups [2]. This crisis of faltering academic achievement in an increasingly competitive world market has captivated a national audience of educators and politicians looking for solutions to problems in the American "pipeline" of mathematics, science, and engineering [21, 26, 33].

Attention among the mathematics community to calculus reform has brought into sharper focus the problem of minorities in the mathematical sciences. As a result, a variety of intervention programs which target underrepresented groups, have been created or continued at most institutions. The American Association for the Advancement of Science (AAAS) has identified twelve key characteristics of academic-based intervention programs. These include:

(1) Academic component focused on enrichment rather than remediation
(2) Highly competent teachers

This project was supported in part by the National Science Foundation, NSF Grant No. MDR-9150212. Opinions expressed are those of the author and not necessarily those of the Foundation.

(3) Integrative approach to teaching

(4) Multiyear involvement with students

(5) Strong leadership

(6) Stable, long-term funding base

(7) Recruitment of participants

(8) Development of peer support systems

(9) Role models

(10) Student commitment to "hard work"

(11) "Mainstreaming" of program elements into the institutional programs

(12) Evaluation, careful data collection, and long-term followup (reported in [20], p. 28).

Among the most widely recognized intervention programs in college mathematics is the calculus workshop model developed for African- American students at the University of California, Berkeley, by Uri Treisman in the late 1970's [3, 14, 31, 37]. The Berkeley model, known now as the Emerging Scholars Program (ESP), has been adapted in mathematics courses at more than a dozen major universities across the country [31], p. 1201, with more than 100 two-year and four-year colleges initiating trial ESP-type programs in the past five years. In a recent issue examining pipeline issues for minority students in Mathematics, Science, or Engineering (MSE) majors, Science magazine (Nov. 13, 1992) reported that some ESP programs have dramatically lowered drop rates and increased the number of minority students majoring in MSE fields. For example, the graduate program in applied mathematics at Rice University and the undergraduate mathematics program at the University of Texas, Austin, each award approximately one-fourth of their degrees to black or Latino students [31]. In 1992, Treisman received a MacArthur "genius" award for this work and its national impact on the success of underrepresented minority students in mathematics-based fields.

Research Questions

The report by Treisman [37] showed dramatic gains in the achievement of African-American workshop students in first-year calculus. This shift in emphasis from remedial to excellence marked the program as one which had many of the twelve key characteristics described by the AAAS. However, despite the growing popularity of ESP-type programs, there have remained unanswered questions regarding their impact, including:

(1) Are there effects on academic performance past the freshmen year when workshop courses are typically offered?

(2) Does the program merely "skim" the best students, or does program participation affect achievement in the course?

(3) Are the types of academic and social issues addressed by the program relevant only for underrepresented minority students, or for non-minority students as well?

The word "effects" in question 1 suggests a causal relationship underlying an association between workshop participation and academic achievement. While historical studies such as this one are limited in the controls that are available for group comparison and analysis, the present study attempts to examine the effects of workshop participation on achievement by combining quantitative and qualitative information. Question 2 is naturally tied to the issue of association versus causality. This question has at its heart the "nature/nurture" dilemma applied in a context of learning mathematics in post-secondary institutions. Question 3 addresses the issue of selectivity, not by the student, but by the institution, in deciding which students or group of students have the highest potential for success or greatest need for help. Implicit in this selection is the choice of measures by which selection is made, whether it be race or ethnicity, gender, or pre-college academic performance. The purpose of this study is to address these three questions in the context of student achievement in mathematics courses and mathematics-based disciplines. While global answers probably do not exist, a case study of one university's experience may help to shed some light on the teaching and learning of mathematics without casting shadows on the teachers and learners.

Theoretical Framework

Theories of academic and social integration have been far more successful in explaining group differences in mathematics achievement than have explanations citing physiological differences. Prominent among the psychologically-based theories is the work of Alfred Bandura [7]. In this seminal work, Bandura described a theoretical model in which self- efficacy expectation, that is, the individual's beliefs regarding her or his ability to successfully perform a given behavior or task, is shaped by the individual's interpretations of past, present, and future experience. More recently, theories of involvement suggest that the individual's academic self-concept and performance remain to a large degree a function of her or his group identification and involvement with socializing agents within the institution [4, 22, 34]. Specifically, in college or university settings, the level of academic commitment and achievement that an individual maintains is affected to varying degrees by the level of her or his group identification. In this sense, academic integration refers to academic achievement, involvement with the curriculum, and interactions with institutional representatives such as faculty and advisors, while social integration includes activities with peers, on-campus activities, and off- campus activities associated with the institution [22, 32]. In the Tinto model [34], this sense of group inclusion is generated in both formal and informal settings, such as structured extracurricular activities and group study sessions or impromptu but significant conversations with peers and faculty. Tinto further suggests that although group involvement is necessary for most students, it is not a sufficient condition. The student must perceive her or his group to be a central rather than marginal part of the institutional structure.

Based on the premise that the purpose of higher education is one of developing student talent, Astin summarizes his theory succinctly: *"Students learn by becoming involved"* ([**4**], p. 133, italics in original). Astin explains this compact idea more fully:

> What I mean by involvement is neither mysterious nor esoteric. Quite simply, student involvement refers to the amount of physical and psychological energy that the student devotes to the academic experience. A highly involved student is one who, for example, devotes considerable energy to studying, spends a lot of time on campus, participates actively in student organizations, and interacts frequently with faculty members and other students. Conversely, an uninvolved student may neglect studies, spend little time on campus, abstain from extracurricular activities, and have little contact with faculty members or other students. ([**4**], p. 134)

Astin sees in this theory the Freudian concept of cathexis, the psychological energy that people invest in objects outside themselves. The construct of involvement is not dissimilar from that described by the phrase "time on task." Astin identifies five basic postulates in the theory of involvement : (1) involvement refers to the investment of physical and psychological energy in "objects," which may be highly generalized (the college experience) or highly specific (preparing for a calculus exam); (2) involvement is a continuous rather than discrete concept - different students will invest various amounts of energy at different times; (3) involvement has both quantitative and qualitative components - for example, the number of hours spent studying for a certain course, versus the quality or effectiveness of that study time; (4) the amount of learning and individual development is directly proportional to the quality and quantity of involvement; and (5) the educational effectiveness of any policy or practice is primarily measured by its capacity to increase student involvement ([**4**], pp. 135–36). For Astin, the important factor for success is the psychological and physiological energy that the student invests in her or his college experience.

Smith [**32**] has described the growing significance of academic integration and social integration for racial and gender groups within increasingly pluralistic campus environments. There is evidence, however, that social integration alone is not effective without a meaningful academic component [**4, 6, 22**]. For some students the campus remains a "chilly" place, especially for women [**27, 28**]. Moreover, some studies have identified a negative impact of disciplines and courses structured on competition rather than cooperation [**5, 10, 27**]. The literature on diversity suggests that perceptions that non-majority students develop about their own academic worth and social fitness are forged by institutional structure, departmental practices, and faculty attitudes. These structural factors that aid or impede student involvement may also deeply affect the process of learning.

Constructivist theories of learning state that knowledge is created or constructed as the individual observes the results of her or his own actions upon objects [24]. Constructivist theory applied to mathematics learning has been extended and tested in several recent reports [15, 16, 17, 25] and books on problem solving [29]. Dubinsky [12, 13] has discussed the implications of the constructivist of learning for restructuring the traditional mathematics classroom from a student passive-teacher active format to a student active-teacher active dynamic:

> If our goal is for students to learn Calculus, then, *what is taught and how it is taught is of no importance whatsoever.* In the last analysis, the only thing that really matters is *what is learned and how that happened—or didn't happen.* ([12], p. 1, italics in original).

Selden and Selden [30] have identified four views on how students learn mathematics: spontaneously, inductively, constructively, and pragmatically. If students learn spontaneously, then the primary role of the teacher is that of lecturer. If students learn inductively, the teacher becomes the provider of materials and problems that allow individual inference. If students learn via mental constructions, the teacher's role is that of facilitator for discussion of and interaction with mathematical ideas. If students learn pragmatically as a response to real world problems, the teacher is the channel through which such problems are accessible. The problem comes when mathematics instructors say they believe students learn through individual interaction with mathematical ideas, but teach as if students learned spontaneously only through the senses of listening and watching ([12], cited in [30]). The college impact models presented here suggest that academic involvement and social interaction produces positive growth, while non-involvement causes decreased interest or disaffection resulting in little or even regressive change. Unlike the developmental theories, though, the college impact models maintain that the institutional characteristics and environment also have an influence on the outcome of student development. In this sense there is a symbiotic relationship between individual involvement and institutional accessibility. Throughout the academic experience, and particularly during the first two years, the individual's academic self- perception is continually evolving and changing to shape the success or failure that ultimately occurs.

The close relationship between cognitive and personal development makes it difficult to isolate specific effects or components of change. The present study tests the general construct of academic and social involvement as a significant agent for change among students enrolled in mathematics-based disciplines. The extent to which factors such as ethnicity or race, gender, or prior academic achievement may affect these students' university career performance is discussed.

Program Description

Upon acceptance to the College of Science or Engineering, each black and Latino student received a letter and personal telephone call from a faculty member or student workshop leader inviting her or him to attend an informational meeting explaining the Academic Excellence Workshop Program. The program co-director indicated that approximately half of the students who were contacted chose to participate in a workshop session for one, two, or three quarters of calculus, while the rest declined. White and Asian students were not eligible to participate in the program, although several students were allowed to join as "guests." The present study included only underrepresented minority students participating in the workshop for one or more quarters of calculus as belonging to the workshop group.

The structure of the workshops was similar to that of the Berkeley program. Each student was enrolled in a traditional lecture section of calculus that included workshop and non-workshop students of all ethnic groups. Unlike the large lectures at Berkeley, the classes averaged about 35 students and met for four hours per week. Since there was no recitation section attached to the course, student questions regarding homework problems and the like were covered in class by the instructor. Workshop and non-workshop students were responsible for the same classwork, homework, and examinations. Although the instructors knew which students were in the workshop, most instructors were "blinded" as to student identity when grading assignments or exams.

In addition to attending the lecture class, workshop students met in structured groups of 10–12 students twice a week for two-hour sessions outside of class to work collaboratively on calculus problems. Group leaders, or facilitators, comprised mainly of upper-division minority undergraduate MSE students, directed the problem-solving activities by constructing worksheets with calculus problems that helped reinforce concepts or expose weaknesses in the students' levels of understanding (see [37], pp. 42–44; [9], p. 279). The facilitators met regularly with the course instructors and workshop director to ensure the relevance of the material presented in workshop sessions and to discuss specific academic problems of individual students. Facilitators would frequently work together to develop interesting and challenging problems. The expectation, which was made clear to workshop students, was that they would excel in, rather than just get through, the course. A typical worksheet problem reviewing exponential growth is given below:

Agronomists use the assumption that 1/4 acre of land is required to provide food for one person and estimate that there are 10 billion acres of tillable land in the world, so a maximum of 40 billion people can be sustained if no other food source is available. The world population in the beginning of 1980 was approximately 4.5 billion. Assuming the population increases at a rate of 2% per year, when will the maximum sustainable population be reached?

During the first half hour of the workshop session, students typically worked alone. Gradually, they would start discussing the problems and comparing solutions. After an hour, the sessions were quite lively, with students explaining their solutions and interpretations to one another. The facilitator then could help individual students or direct the discussion of group questions that arose. The course instructor wrote the exams used in her or his course, so the facilitator had no information relevant to exam questions. The sessions also had an informal social aspect, with students sometimes munching on popcorn or pizza while they worked. Moreover, the discussions often included non-mathematical topics as well, such as information on future course sign-ups or deadlines, upcoming departmental activities, or personal concerns.

Method

Sample. The present research was a longitudinal study of the academic performance of minority and non-minority students enrolled in a mathematics-based major at the College of Engineering or the College of Science at California Polytechnic State University, Pomona (Cal Poly). The sample was comprised of 133 Latino and African-American students who participated in at least one calculus workshop section while enrolled in a traditional lecture calculus sections. The performance of this group was compared to that of three peer groups of students enrolled in the same lecture sections of first-quarter calculus. These included 187 African- American and Hispanic students not enrolled in the workshop, 208 white, non-Hispanic students, and 198 Asian or Pacific Island students, a total sample of 726 students. The workshop sample was comprised primarily of Latino students (116/133), while about one-fourth (36/133) of the workshop students were women. Similarly, the non-workshop minority group was comprised mostly of Latino students (159/187), including 21% women (39/187). Both the white and Asian groups had a similar gender mix of 30% (62/208) and 34% (67/198) women, respectively.

Instruments. Three types of measures of students' performance were taken. The first measure was an analysis of each student's academic performance over a period of three to five years from Fall, 1986, through Spring, 1991, depending on the student's year of entrance. Mathematics course grades (A = 4.00), including the number of times a student repeated a course because of failure, were taken directly from student transcripts of grades at Cal Poly. A Course Attempt Ratio (CAR), the mean number of quarters a group of students needed to successfully complete the first-year calculus sequence, provided a secondary measure of student achievement and cost effectiveness. For example, if five students successfully completed the three-quarter sequence in, respectively, 3, 3, 4, 4, and 7 quarters, the CAR for that group is 4.2. Finally, each student was evaluated on whether he or she was still enrolled or graduated in MSE after three years, had left MSE but was enrolled in a non-technical discipline, or had left the institution entirely.

The second measure was a survey instrument similar to that developed by Pascarella [22]. First-year calculus students completed the Student Involvement Questionnaire regarding time spent in weekly course preparation, outside commitments, and institutional or departmental involvement. The main purpose of this tool was to determine if workshop participation was linked to differences in outside commitments or time spent in individual or group study. The third measure was an interview with upper-division underrepresented minority students who had participated in the workshop as freshmen regarding their perception of how or if the workshops had affected their academic lives. Each student was asked the same set of nine questions, with follow-up questions to help clarify or augment a response (Appendix A).

Results

Pre-College Achievement Measures. The academic preparation of workshop and non-workshop students was evaluated to examine which groups of students, if any, may have had a stronger academic background prior to entering the institution. These measures included SAT-Math, SAT-Verbal, High School Grade Point Average, and the student's score on a pre-calculus Placement Exam given by the mathematics department. Table 1 reports the mean scores on these measures for each group of students.

TABLE 1. Pre-College Measures and Mathematics
Achievement of MSE Students, 1986–1990

	Black and Latino		Asian	White
	Workshop	Non-Workshop	Asian	White
n	133	187	198	208
Pre-College Academic Measures*				
SAT-Math	525	515	579	611
SAT-Verbal	405	415	394	487
High School GPA	3.32	3.23	3.38	3.37
Diagnostic Test	29.0	29.4	34.3	32.1
Mathematics Achievement				
Mean Grade in First-Quarter Calculus	2.69	1.76	2.40	1.98
Enrolled in MSE After 3 Years	85%	48%	59%	50%
Mean Grade in First Two Years of Calculus	2.40	1.74	2.50	2.38
Course Attempt Ratio in First-Year Calculus	3.63	4.64	3.67	3.77

* Some measures not available for all students

There were no statistically significant differences between the minority workshop and minority non-workshop groups in any of the four pre- college academic measures. Likewise, there were no differences between white and Asian students in SAT-M, HSGPA, or Placement Exam, although the Asian students scored lower in SAT-V ($t = 3.42$, $p < .01$). Comparing black and Latino students as a group to white and Asian students as a group showed two differences. The black and Latino group of students scored significantly lower on SAT-M ($t = 2.38$, $p < .01$) and the Placement Exam ($t = 2.34$, $p < .01$) than did the white and Asian group. High school GPA's were almost identical for all groups, with a mean of 3.33 for all students.

The above comparisons indicate that the pre-college achievement measures for black and Latino students in the workshop were no different than those for non-workshop black and Latino students. Thus, for this sample, students' selection into the Academic Excellence Workshop program was not associated with pre-college achievement. However, the underrepresented minority students began their calculus sequence with lower measures of mathematics achievement than did their non-minority peers. Interpreting high school GPA as a measure of exposure suggests that none of the four groups had an initial advantage in adapting to the rigors of a university-level calculus course. The differences found here indicate that the workshop group of black and Latino students had no advantage in mathematics background or achievement over their peers at the start of the calculus sequence.

Comparison Between Minority Groups. In a comparison between minority groups, workshop students achieved a mean grade of more than six tenths of a grade point above non-workshop students in first-year and second-year calculus ($t = 3.75$, $p < .001$). Moreover, within three years after entering the institution, more than half (52%) of the minority non-workshop students had either withdrawn from the institution or changed to a non mathematics-based major, compared to fifteen percent of the workshop students (Table 1). Computation of Course Attempt Ratios showed that non-workshop students required an average of one full quarter more to complete their three-quarter calculus sequence due to course failure. Individual records showed high patterns of course-repeating, with nearly half (46%) of the non-workshop minority students requiring 5 or more quarters to complete a 3-quarter calculus sequence, compared to fewer than one-fifth (17%) of the workshop students. Moreover, 91% of the workshop students still enrolled in MSE majors after three years had completed their mathematics requirement in their respective MSE majors, compared to 58% of the non-workshop minority students.

The association of workshop participation with academic success and persistence was particularly strong among minority women. All of the women (22/22) who had taken the calculus workshop were still enrolled at Cal Poly after three years, with 86% (19/22) remaining in a mathematics-based major. Of these, each had completed her mathematics requirement. Conversely, 52% of the original group (12/23) of non- workshop minority women were still enrolled after three years, with only four of these women persisting in a technical major and successfully completing her mathematics requirement.

Comparison With Non-Minority Groups. In the first-year and second-year calculus sequence, workshop black and Latino students achieved the same as white students and slightly (but not significantly) lower than Asian students (Table 1). After three years, 50% of the white students and 41% of the Asian students had withdrawn from the MSE major or left the institution, compared to 15% of the workshop students. The pattern of course repeating was almost identical between non-minority groups and workshop students, with each group averaging about 3.7 quarters to complete the three-quarter sequence. Eighty-two percent of the white students and 89% of the Asian students had completed their mathematics requirement within three years, proportions that were nearly identical to the 91% completion rate for workshop black and Latino students. Controlling for gender showed that white and Asian women had MSE attrition rates of 38% and 45%, respectively, while those for white and Asian men were 55% and 39%, respectively.

Student Involvement Questionnaire. The 1990-91 cohort of students currently enrolled in first-year calculus when this study was conducted completed the Student Involvement Questionnaire [22] to assess study patterns and outside involvement. Black and Latino workshop students reported studying a total of 8.2 hours per week for their calculus class, about 1.3 hours per week more than for white and non-workshop minority students, and 0.5 more than for Asian students. However, four of these hours were spent in group study in the workshop sessions, compared to an average of one hour spent in group study for all other non-workshop students. Black and Latino women participating in the workshop reported the most weekly study time of all subgroups, spending more than ten hours per week outside of class studying calculus. The data do not suggest, though, that this group of workshop students spent more time in individual study than did the non-workshop students. Similarly, there were no significant differences between ethnic or gender groups in the amount of time involved in on-campus or off-campus activities.

Workshop Student Interviews. Personal interviews were conducted by the author with 23 of the 45 former workshop students from the 1987 and 1988 cohorts regarding their perceptions of their workshop experience. Each of the 45 students was sent a letter explaining the purpose of the study and a request for an interview. Followup telephone calls were made by the author to each of the 45 students. Conversations were eventually had with twenty-five of the students, all but two of whom agreed to meet for an interview. Twenty-one of the students arranged a personal interview in an office on- campus, with two of the interviews taken by telephone because of scheduling conflicts. Interviews ranged from 15 to 60 minutes in length, with the majority lasting around 30 minutes. The sample was comprised of 22 Hispanic students, including 9 women and one African-American man. Nineteen of the 21 in-person interviews were tape-recorded with written permission of the student. The interviews were then transcribed by the author with the responses grouped into one of three basic types: agree, disagree, or neutral. There were, of course, gradations between these, but for simplicity and clarity, each response was evaluated as essentially one of these three.

The primary academic impact reported by interviewees was that workshop participation had "brought them up to the level" of content mastery that would be expected of them on calculus examinations in their first year of college. Five students discussed their amazement at being academically challenged by the other workshop students. M. J., a junior in Industrial Engineering, said that he was a

> ... good student in high school, in the top 5% of my class. But the workshop people were smarter than me or more disciplined. I saw that people were better than me. They challenged me to get to their level.

More than half of the students interviewed mentioned that while they had difficulty making time to study for calculus, the accountability of the workshop sessions motivated them to study. R. D., a student majoring in Mechanical Engineering, articulated her realization of the amount of study time necessary to succeed in technical courses:

> You realize how much time you put into that one class, and that was, you know, a freshman-sophomore level calculus, so you get higher, you think 'Gosh, it took me that much time, plus the workshop, plus the tutoring,' it makes you really realize that you do need to put in time, it takes time to do your classes.

Hence, although workshop effects on individual study time were mixed, the majority of students reported that their study time had changed both in terms of quality and quantity.

However, not all workshop participants desired this level of academic involvement. One interviewee who had dropped out of the workshop indicated that the program was good for "serious" students, but required too large of a time and energy commitment. Although he agreed in theory that group learning was helpful, he did not perceive the need for personal involvement:

> I can do it on my own. I work 35 hours per week. But sometimes I didn't work I would use that as an excuse. If I did have time, I would waste it, not work on school.

Thus, although his work schedule made workshop participation difficult, he admitted that even without work he probably would not have felt much differently. Three students who had dropped out of the workshop indicated that the workshops were too demanding and required too much extra time. While finding time for the four hours each week was a concern for some students interviewed, the majority of these students indicated that it was more a matter of making time rather than having time. Although more than half of the students had some type of part-time employment off-campus, they felt that the positive effects of attending workshop sessions outweighed the personal inconvenience or hardship of scheduling conflicts with work or other commitments.

The degree to which the workshop was perceived to be helpful did not appear to be the same for the men and women interviewed. While eight of the nine women felt that they would not have done as well in their calculus classes without the workshop, more than one-fourth (4/14) of the men indicated that they probably would have done as well either way. Most (18/23) of the students indicated that they probably did not spend more time in individual study while they were in the workshop courses, with several students reporting that they could actually spend less time in individual study and still get what they needed during the workshop sessions. Thus, students indicated that the quality of the workshop sessions (types of problems, difficulty of problems, talking with peers, practice exams) was at least as important as the quantity of time spent.

Interview students were asked if participation in the workshop had affected their decision to continue in their MSE major. More than half (5/9) of the women reported a direct effect, compared to less than a third (4/14) of the men. The common theme among the nine students indicating a positive effect was that of encouragement. In this study, peer encouragement seemed to be especially important to the women. This encouragement sometimes came in the form of discipline and accountability. O. S., a senior majoring in Computer Science, reflected about her first year:

> I never thought I could do it. I had to take remedial courses courses, couldn't pass the stupid diagnostic test. I got a B+ in physics. Before it was, like, C's. It brought me up one to two levels from F to C, or D to B.
>
> [The facilitators] push you to a limit that you think you're gonna die, that you can't do it any more, and they say, 'It's not that hard.' I had many facilitators stay with me until I got that problem, 'cause they knew I wouldn't do it if I left.

S. B., a senior in Electrical and Computer Engineering, mentioned the importance for her of the facilitator as a role model:

> I remember one specific math facilitator. She was Tau Beta Pi, the top 20% for engineers. You look up to them.

However, several of the men indicated the importance of encouragement of their peers. M. J., a junior in Industrial Engineering, summarized the importance of encouragement after a test:

> It's the whole thing about the workshop, you know, to get you up
> there to another level. It's discouraging sometimes, you do bad on a
> test, you go to the workshop, it's like, 'Man, how you doing?' 'Not
> too good.' 'Let's talk about it. What was the problem?' [Your peers]
> encouraged you a lot.

Conversely, about two-thirds of the students felt that the workshop had not
guided their decision to stay within a technical field. However, several students
stated that working in an environment where excellence was expected confirmed
their sense of being potential engineers or scientists. Thus, participation in the
workshop had little impact on the student's decision to remain in a mathematics-
based major. However, mutual encouragement and accountability from peers and
the workshop facilitator had a direct effect on maintaining interest in and success
in the calculus course, particularly among women.

Whereas half (7/14) of the men indicated that they had continued to work
in groups in their subsequent courses, all nine of the women reported not only
forming small groups on their own, but choosing classes and schedules so as to
be together with other female black and Latina students who were successful
academically. Thus, while workshop participation had an effect on the academic
behavior of some of the men in their subsequent courses, it seemed to impact
all of the women, both academically and socially, a finding consistent with other
gender studies [**10, 18, 27, 28**]. Both men and women students reported that
although the notion of being in an "excellence program" was appealing, the
reason they joined initially was because they were personally invited and felt
like part of the group.

One half of the Latino students interviewed expressed strong views on the
racial or ethnic mix of the workshop program. Since participation was limited to
African-American and Latino students, several students reported stronger ethnic
identity in regard to mathematics-based fields as a result of seeing their ethnic
peers excel. However, at least one student reported a strong resentment of the
exclusive nature of the program:

> Why is it just for us? Do we really need that much more help? In
> [previous institution], the workshop was open to all groups. I was
> uncomfortable that the workshop was for minorities only. It felt like
> cultural singling out.

The nine women who were interviewed were asked if their workshop experience
had affected their perception of women in technical fields. Each of the students
indicated that it had, with several of the students having a strong reaction to
the question. The common theme was that of isolation in a male-dominated
discipline. C. T., a senior in Electrical and Computer Engineering, summarized

the problem:

> There are very few women in ECE classes. You look around, see half
> Asian, half white, and you're the only one. It's kind of lonely.

O.S., a Computer Science major, implied that for her, the issue of gender was
far more important than ethnicity or race:

> There are very few women in Computer Science and in technical fields.
> Being a minority, Hispanic, and being female, when you're walking
> to class, you're the only girl, and it was very awkward. It doesn't
> matter if you're Hispanic, or orange, or purple, or whatever. It's like
> the guys are, 'What is she doing here?', that kind of attitude. They
> get over it, 'cause ... you're doing good in the class. When you are
> doing better [than them], it's a shocker!

Moreover, C. L., an above-average student majoring in Mechanical Engineering,
indicated the importance of having met "a lot" of female engineers. Although she
was glad to have had two workshop facilitators that were women, she discussed
the barriers to women in engineering:

> You hear about how few women actually graduate in engineering,
> even the ones that start in it, and I seem to think it might be harder
> for women to actually graduate in it. Whether they mean to or not,
> you see a lot of, you know, prejudice against the women ... within
> the teachers or fellow students as if they don't seem to think you can
> actually make it sometimes.

She concluded with the importance of women role models who have completed
the engineering degree:

> Just seeing that other women have made it, you feel that maybe you
> can make it.

While the experiences and perceptions of the 23 workshop students interviewed
may not have been representative, they reported having an experience that was
academically powerful and personally meaningful, especially for women students.

Discussion

Question 1: Subsequent Workshop Effects. This study found that African-American and Latino students who participated in the workshop calculus
sessions achieved as high or higher than any other ethnic group of students,
both during the first-year and beyond. Figure 1 shows a boxplot of mean grades
of first-year and second-year calculus courses for each student persisting in an
MSE major.

The median calculus GPA for each group given in Table 1 showed no differences between white, Asian, and workshop minority students at the 25th or 50th
percentiles. However, comparison between groups at the 75th percentile showed

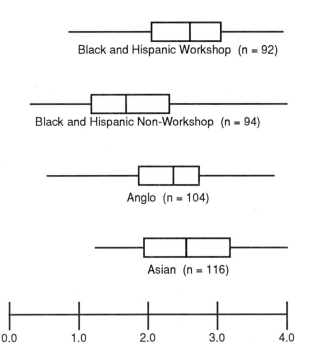

FIGURE 1.   Boxplot of Mean First-Year and Second-Year Calculus
Grades, 1986–1989

Asian students above 3.00, with workshop minority and white students at 2.83
and 2.72, respectively.

It can not be determined from this study all of the factors that may have contributed to this measure of student outcome. The self-selectivity of the workshop
students may in part explain the wide difference between the median calculus
GPA of minority workshop (median = 2.40) and non- workshop (median = 1.74)
students. In this sense, though, the white and Asian students also are select
groups in that they persisted in the MSE major. A comparison of SAT-M,
SAT-V, HSGPA, and placement test scores between each of the four groups of
persisting students showed no statistically significant differences between persisting and non-persisting students in any of the four measures ($t < 1.00$, $p > .15$).
Moreover, comparing pre-college achievement measures between ethnic groups
of persisting students showed no differences other than those reported earlier for
the initial sample. Thus, while social or personal factors contributing to student
achievement undoubtedly impacted student achievement and persistence, this
study found that pre-college measures of achievement were not associated with
student success [34].

The interview results suggest that many workshop students, especially women,
felt that the workshop was a critical part of their academic and social develop-

ment in college. Indeed, most of the women students interviewed had a difficult time separating these experiences. While the interview group may not be representative of workshop students in general, there were positive academic effects associated with the academic and social involvement associated with group participation. This finding is consistent with Astin's research on the effects of cooperative environments in higher education [5], here applied in the setting of first-year calculus.

Question 2: Skimming and Self-Selection. The critical issue of self-selectivity cannot be fully addressed in an historical study. Psychological measures of personality, professional drive, personal circumstances, and other "non-academic" aspects not addressed here would greatly help remove some of the shadows. However, this study did find some evidence to indicate that calculus achievement for workshop students was, at least in part, a result of developing student talent rather that merely selecting the most talented students.

The grades of workshop minority students enrolled in first-quarter fall calculus sections from 1986 to 1990 were compared historically with those of all other students enrolled during this time period, as well as with students enrolled during the pre-workshop era from 1980 to 1985, a total sample of 5,315 students over a period of 11 years (Figure 2). Since departmental course grade point averages varied considerably over this time period (high = 2.38, low = 1.56), z-scores rather than course grades were used for comparisons over time.

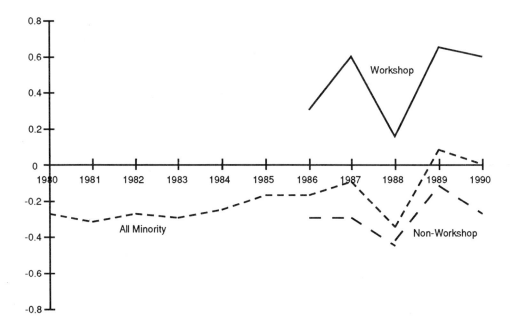

FIGURE 2.   Z-Scores of Calculus I Grades, 1980–1990

Figure 2 shows that from 1980 to 1985, mean grades of first-quarter calculus black and Hispanic students ranged from .34 to .14 standard deviations below departmental means; the weighted average during this time was −.28. During the workshop era of 1986–1990, mean grades of all black and Hispanic students, including those in and those not in the workshop, ranged from −.34 to 0, a weighted average of −.14. Thus, while grades still varied from year to year (especially in 1988), the first-quarter calculus performance of the population of underrepresented minority students at Cal Poly measurably increased after the workshop program began.

The performance of non-workshop minority students from 1986–90 was compared with that of all minority students prior to 1986 (Figure 2). As reported above, the mean calculus z-score for all black and Latino students was −.28. From 1986 to 1990, the weighted mean score for non- workshop minority students was −.24, while that for workshop students during this time was .50. Thus, from 1986 to 1990, minority students in the workshop achieved five tenths of a standard deviation above departmental means, while the achievement of non-workshop minority students remained at the same level as during the pre-workshop era. This strengthening of the calculus performance associated with the workshop program for the population of primarily Hispanic students replicates a result reported by Treisman for a population of African-American students [37]. The results in both studies are consistent with a causal relationship. However, the extent to which workshop self-selection is tied to other measures of college achievement such as persistence, graduation rates, and subsequent professional success is an area for further inquiry.

Question 3: A Minority Issue? Reports by the National Research Council [19, 20] and *Science* magazine [31] have helped increase awareness of the disproportional failure of underrepresented groups of students, including ethnic, gender, age, and disability groups. This study found some evidence of feelings of ethnic separation and academic inadequacy among those minority students interviewed despite their success in the program. Although the workshop program was not necessarily designed to address gender issues, the most compelling results were those for women in the workshop. Although based on a limited sample ($n = 22$) of Hispanic women, this group had a persistence rate in the university of 100% and in their MSE majors of 86%. The nine women who were interviewed indicated a direct effect of workshop participation with their success, both in study patterns and peer relationships. Moreover, feelings of exclusion or inadequacy seemed to be centered on gender rather than ethnic issues. The relatively high attrition rates of non-minority women (more than 40%) may indicate a similar pattern of gender separation in their technical courses.

Summary. The data presented here may be merely the tip of a much larger iceberg of student failure. In *The Challenge of Diversity*, Daryl Smith states that many non-minority students are underprepared and underachieving in college [32], p. 8. The experience reported by some of the underrepresented minority women, together with the unprecedented success of women who were in the workshop, suggest that intervention models which promote academic excellence through intentionally structured groups may be especially effective among women [28]. However, this study found the highest attrition rate (55%) among white men majoring in mathematics-based disciplines. It is reasonable to expect that participation in an excellence workshop program would have some effect on these (or other) students participating. However, it is not clear the extent to which factors behind student success or failure cut across gender and ethnic lines.

Impact models of student development [4, 22, 34] have identified subtle but effective ways in which an institution can assimilate or marginalize groups of students. The present study found that the workshop seemed to provide both a necessary and sufficient social structure for many participants, and especially for women, to succeed. Since the workshop program did not include non-minority students, the present study was unable to test possible program effects on these students, and how these effects might vary by ethnic group. Experimentation in program structures could range from high exclusivity to mandatory inclusion. In either case, the goals and expectations for such a program should be made clear to all of those involved. It is doubtful that either extreme would effectively address many of the twelve key characteristics of academic-based intervention programs identified by the American Association for the Advancement of Science [20].

Conclusion

The present study found evidence that an intervention program promoting academic excellence and peer interaction in an academic context can directly affect student performance in technical majors independently of pre-intervention cognitive factors. Methodological problems regarding sampling, randomization, and representativeness preclude global generalizations of the case study presented here. However, three findings emerged that may hold significance for change of traditional instructional practices at colleges and universities. First, the common pattern of course failure and repetition raises the issue of financial costs associated with student failure. Cost analysis indicated that the cost associated with a student's repetition of classes was far greater than the per-student cost of the workshop [9], pp. 235–240. For example, during the academic year 1990-1991, the per-student cost to the state for a four-unit calculus course was $517 per quarter, while the per-student cost for the workshop program was $335 per year. Since workshop minority students completed their three-quarter sequence in one quarter less than did non-workshop minority students (Table 1), the cost for a student to repeat a course outweighed that of placing the student in the workshop for that year. This logic assumes, of course, a causal relationship between workshop participation and calculus achievement which may not be justified. Also, structural differences between funding for FTE's (full-time enrolled students) and special intervention programs make it difficult to assess who is paying and how much. However, close examination of the reality of course repeating raises deep questions and concerns about the costs of student failure, particularly at state-supported institutions. Non-minority students were not immune from this pattern of recycling through courses: transcript searches found that one white male student took 15 attempts (without withdrawing) to complete six quarters of calculus and linear algebra.

Second, the present study's use of hand-tracking the progress of individual students underscores the importance of having computer systems which can accurately track students' progress across semesters, including course repetitions and withdrawals, in order for academic departments or divisions to measure students' real progress and the impact of particular programs or courses. And third, while the study could not clinically control for motivational factors in workshop self-selection, the study found the existence of a non-trivial number of students who seemed to use the program effectively and, as a group, achieved unprecedented success in mathematics-based majors.

The extent to which ESP-type programs are transferable to other levels or disciplines remains unclear. The self-selection rate of around 50% indicates that there may be a significant number of students who would participate in workshop-type programs at other institutions. However, the Cal Poly workshop program, a joint effort of mathematics, science, and engineering faculty and administrators, was viewed as an important part of these departments' work. The program required a high level of time commitment from its co-directors, who developed a rigorous training program for workshop facilitators, and who closely monitored their work and their students' progress [1]. The findings here indicate that using non-lecture methods of teaching and group learning may have a significant impact on the performance and involvement of some students. In summary, the data strongly suggest that achievement among underrepresented minority students in mathematics, science, and engineering disciplines may be less associated with pre-college ability than with in-college academic experiences and expectations.

Appendix A. Interview Questions for Workshop Students

(1) Do you think you would have done as well in your calculus course if you had not been in a workshop?
(2) Did participation in the workshop affect the amount of time you spent outside the workshop preparing for the course?
(3) Do you think there has been any effect of the workshops on your study habits or study methods since that time?
(4) Do you feel that working together in groups was helpful in learning the material?
(5) Did the workshops help you in your transition to the campus? If so, how?
(6) Did the workshop experience affect your decision to stay in or withdraw from a math or science major?

(7) Has the workshop experience affected your perception of Latinos or blacks in mathematics, science, and engineering?

(8) Has the workshop experience affected your perception of women in mathematics, science, and engineering?

(9) For you, what was the most important aspect of your workshop experience?

REFERENCES

1. Academic Excellence Workshops, *A handbook for academic excellence workshops*, Minority Engineering Program and Science Educational Enhancement Services, Pomona, CA, 1991.

2. N. Alsalam (ed.), *The condition of education, vol. 2*, National Center for Education Statistics, Washington, D.C., 1991.

3. American Mathematical Society, *Research mathematics in mathematics education*, AMS Notices **35** (1988), 1123–1131.

4. A. Astin, *Achieving educational excellence*, Jossey-Bass, San Francisco, CA, 1985.

5. _____, *Competition or cooperation?*, Change **19** (1987), 12–19.

6. _____, *What matters in college? Four critical years revisited*, Jossey-Bass, San Francisco, CA, 1993.

7. A. Bandura, *Self-efficacy: Toward a unifying theory of behavioral change*, Psychological Review **84** (1977), 191–215.

8. M. Berger, *Predicted future earnings and choice of college major*, Industrial and Labor Relations Review **41(3)** (1988), 418–429.

9. M. Bonsangue, *The effects of calculus workshop groups on minority achievement and persistence in mathematics, science,and engineering*, Doctoral dissertation, Claremont, CA, 1992.

10. B. Clinchy, M. Belenky, N. Goldberger, N., and J. Tarule, *Connected education for women*, Journal of Education **167** (1985), 28–45.

11. J. Dossey, I. Mullis, M. Lindquist, and D. Chambers, *The mathematics report card: Are we measuring up? (Report No.17–M–01)*, Educational Testing Service, Princeton, New Jersey, 1986.

12. E. Dubinsky, *A learning theory approach to calculus*, Proceedings of the St. Olaf Conference on Calculus, Oct. 20-22, 1989.

13. _____, *Reflective abstraction in advanced mathematical thinking*, Epistemological foundations of mathematical experience (L. P. Steffe, ed.), Springer-Verlag, New York, 1990.

14. R. Fullilove and P. U. Treisman, *Mathematics achievement among African American undergraduates at the University of California, Berkeley: An evaluation of the mathematics workshop program*, Journal of Negro Education **59** (1990), 463–478.

15. C. Hoyles, *What is the point of group discussion in mathematics?*, Educational Studies in Mathematics **16** (1985), 205–214.

16. J. Kilpatrick, *Reflection and recursion*, Educational Studies in Mathematics **16** (1985), 1–26.

17. S. Lerman, *Constructivism, mathematics, and mathematics education*, Educational Studies in Mathematics **20** (1989), 211–223.

18. M. Linn and J. Hyde, *Gender, mathematics, and science*, Educational Researcher **18** (1989), 283–287.

19. National Research Council, *Everybody counts: A report to the nation on the future of mathematics education*, National Academy Press, Washington, D. C., 1989.

20. National Research Council, *Moving beyond myths: Revitalizing undergraduate mathematics*, National Academy Press, Washington, D. C., 1990.

21. National Science Foundation, *Women and minorities in science and engineering*, Washington, D. C., 1990.

22. E. Pascarella, *Racial differences in factors associated with bachelor's degree completion: A nine year followup*, Research in Higher Education **23** (1985), 351–373.

23. I. Peterson, *The troubled state of calculus*, Science News **129(14)** (1986), 220–221.

24. J. Piaget, *Genetic epistemology*, Columbia University Press, New York, 1970.

25. S. Pirie and R. Schwarzenberger, *Mathematical discussion and mathematical understanding*, Educational Studies in Mathematics **19** (1988), 459–470.

26. Quality Education for Minorities Project, *Education that works: An action plan for the education of minorities*, Cambridge, Mass, 1990.

27. B. Sandler, *The classroom climate: Still a chilly one for women*, Educating men and women together (C. Lasse, ed.), University of Illinois, Carbondale, 1987.

28. B. Sandler and R. Hall, *The campus climate: A chilly one for women?*, Association of American Colleges, Washington, D. C., 1982.

29. A. Schoenfeld, *Mathematical problem solving*, Academic Press, Orlando, Florida, 1985.

30. A. Selden and J. Selden, *Constructivism in mathematics education: A view of how people learn*, UME Trends **2(1)** (1990), 8.

31. P. Selvin, *Math education: Multiplying the meager numbers*, Science **258** (1992), 1200–1201.

32. D. Smith, *The challenge of diversity: Involvement or alienation in the academy*, The George Washington University, Washington, D. C., 1989.

33. L. Steen (Ed.), *Calculus for a new century: A pump, not a filter*, Mathematical Association of America, Washington D. C., 1987.

34. V. Tinto, *Leaving college*, University of Chicago Press, Chicago, 1987.

35. H. Torsten, *International project for the evaluation of educational achievement*, Wiley, New York, 1967.

36. K. Travers, M. McKnight, and J. Dossey, *Mathematics achievement in U. S. high schools from an international perspective*, NASSP Bulletin **69(484)** (1985), 55–63.

37. P. U. Treisman, *A study of the mathematics performance of black students at the University of California, Berkeley*, Doctoral dissertation, Berkeley, California, 1985.

CALIFORNIA STATE UNIVERSITY, FULLERTON, CA 92634

CBMS Issues in Mathematics Education
Volume 4, 1994

The Case of Dan: Student Construction of a Functional Situation through Visual Attributes

STEPHEN MONK & RICARDO NEMIROVSKY

Introduction

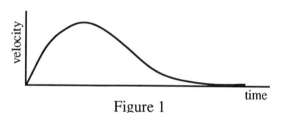

Figure 1

In interpreting the graph of velocity vs. time in Figure 1, an experienced user of graphs might see it as describing an event in which an object moves forward for the entire time interval, starting from a standing position. The object accelerates rapidly, reaches a top speed, and then steadily decelerates to a stop, although at a slower rate than it had speeded up at. In making such an interpretation, this person is responding to the visual information of the graph which she might see as being made up of two segments, one that bends sharply upward and another that is somewhat straight and heads downward, but is less steep than the first. One segment begins on the horizontal axis, and the other ends there; they are joined by a rounded peak between them. This visual information is so intertwined with her understanding of functions, motion, and the relationship among velocity, distance, and time that she is hardly aware of the graph as a visual object. Nonetheless, the overall shape of the graph is a highly compressed and powerful carrier of information for her. While she is responding to the graph

The first author supported by National Science Foundation Grant #MDR–9155746. The second author supported by National Science Foundation Grants #MDR–8855644 and #MDR–9155746. All opinions, findings, conclusions, and recommendations expressed herein are those of the authors and do not necessarily reflect the views of the funder.

as a thing-in-itself, she is also, in the words of Mason ([6] p.74), *seeing through* the graph to a wider field of concepts and relations. Although the graph as a visual object is a crucial aspect of her interpretation, it is largely *transparent* to her.

The literature on students' understanding of graphs contains numerous studies in which students respond to the visual qualities of graphs in what are considered to be inappropriate ways. The results of most of these studies are described in the review article on student understanding of functions and graphs by Leinhardt, Zaslovsky, & Stein [5], where they describe the most extreme form of such a misreading of graphs as one in which students "interpret a graph of a situation as a literal picture of that situation" (p.39). Typical of such a response, which they call an Iconic Interpretation, would be one in which the student acts as if he or she literally sees the graph in Figure 1 as a hill and regards the moving object as going up and then down the hill.

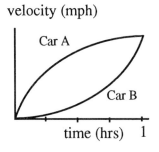

velocity (mph)

time (hrs)   1

Figure 2

In another study in which students misinterpret the visual qualities of a graph, Monk [8] describes a problem on a college calculus examination in which the students were given the velocity vs. time graphs for two cars shown in Figure 2, and were asked to describe where the cars are in relation to each other at various times in the interval shown. Fifty five percent of these students responded by saying that the two cars are coming together at 1 hour, even though they indicated on another question that they knew that the individual points on the graph give velocities of the two cars at various times. Clement [2], McDermott, Rosenquist & vanZee [7], among other authors, report similar results. While these students have not interpreted these graphs as if they were an actual picture of the situation of the two cars moving along, they have made an overly direct connection between its visual features and the situation it represents.

The cumulative effect of such studies is to suggest to many that, even though the shape of a graph is a rich resource for an experienced graph user like the one described above, it is primarily, if not exclusively, a source of confusion and error for students. If a student responds to a graph in such a way as to interpret it either as a picture of the situation it represents or as carrying directly translatable information about the situation, this student would seem to be better off if his or her visual responses to graphs could, at least temporarily, be ignored or suppressed. In this paper we challenge this conclusion by presenting a case study of a student whom we call "Dan," whose use of graphs depends crucially on his response to their visual qualities. Even though Dan's understanding of the functional situations represented by these graphs has its inadequacies, it becomes richer and more adequate throughout the interview. In this study, we give a detailed description and analysis of a clinical interview with Dan in which he interacts with a physical device whose behavior can be regulated and described by functions represented by graphs. Dan proceeds from a visual approach in which he divides a graph into segments to which he ascribes meaning through their visual attributes such as up vs. down, high vs. low, and more or less steep. We want to show in detail how he interacts with a graph, since, while he responds to it as a visual object, he continually refines the way in which he interprets its visual qualities. While he does make overly simple connections between visual features of the graph and the physical context, he is able to move beyond such connections to make other more complex connections between the graph and the physical events. In our examination of this single case we wish to show that, while Dan makes errors and has misunderstandings about the functional situation, he also expands and clarifies his understanding of it.

We believe that this study has implications of several kinds. First, it suggests a changed attitude toward the use of the visual aspects of graphs in the development of students' understanding of functions and graphs. Rather than being merely a source of confusion or a trigger for erroneous conceptions that have to be replaced, the visual aspects of graphs can become a considerable resource in students' learning. Second, the way in which Dan's understanding of this functional situation changes provides further evidence in support of a view of student learning that has recently emerged among educational researchers (see, for example, Smith, diSessa, & Roschelle, [11]): that a description of student conceptual change framed in a language of misconceptions and "erroneous" ideas that must be "replaced" should itself be dropped in favor of a view of learning that takes place through processes of gradual refinement of a complex system of ideas and experiences. Third, the processes through which Dan modifies his understanding by being able to experiment with and communicate about graphical shapes and physical events directly connected to these graphs suggests criteria for the design of learning environments that can promote such learning.

Overview of the study and this paper. This paper is based upon an interview that was part of a series of teaching experiments carried out by Ricardo Nemirovsky and his associates at TERC. Among the aims of these studies is to investigate more deeply the ways in which high school students understand functions, especially those that are connected to physical devices and are represented by graphs, diagrams, and tables. Each interview carried out by this group is based on a single physical device which the student can use to explore and reason about a small family of functions. Each such device also has the capacity to produce, in real time, graphs that result from the student's operation of the device, so that the student can see, in real time, the consequences of his or her constructions of the functional situation.

In the interview upon which this paper is based, the second of a series of three interviews with Dan, he uses a device called the Air Flow Device. (An earlier paper based on this interview with Dan was written by Nemirovsky & Rubin [10].) This device, shown in Figure 3, contains a bellows, which can be used by the subject to push or pull air into or out of an air bag connected to the bellows by a hose. When the bellows is pushed down, the Air Level in the air bag goes up, and when the bellows is pulled up, the Air Level in the air bag goes down. A weight can be been placed on the top of the bellows, so that when it sinks, it pushes air into the air bag, making the Air Level in the air bag go up. In the hose between the bellows and the air bag there is both a valve which can be adjusted to control the Flow Rate of air between the bellows and the air bag and a sensor which monitors the flow of air. The sensor is connected to a computer which generates, for any "event," in which the student has operated on the device for a small amount of time, both a graph of Flow Rate vs. time of the air in the hose, as well as a graph of Volume vs. time of air in the air bag.

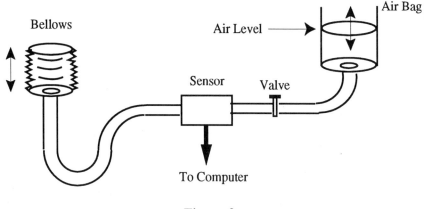

Figure 3

These interviews are both video and audio recorded. There are two video cameras, one which focusses on the subject, the interviewer, and a large pad of paper mounted on an easel they both write on, and a second which is aimed only at the monitor screen of the computer that produces the graphs made from the readings sent by the sensor. Thus, the data used in our discussion of the interview consists not only of the video tape of the conversation between interviewer and subject (including gestures, such as pointing to and tracing graphs), but also copies of the sheets on which they have made various sketches of graphs, as well as the images of graphs made by the computer and recorded on the video tape.

At the time of the interview, Dan was a high school student in the twelfth grade who had taken, or was taking, three years of college prep mathematics as well as a course in physics. The general task set for Dan in these interviews is to begin with a simple graph of Flow Rate vs. time, such as an oblique straight line, and to predict the graph of Volume vs. time that would be associated with it. In each case, Dan is encouraged to proceed by attempting to produce an event with the device which has the given graph as its Flow Rate vs. time graph and then to compare the actual experimental Volume vs. time graph with his prediction. The interview is conducted by Ricardo Nemirovsky, referred to as "RN" in the transcript and our discussion of it.

We proceed with our analysis of how the visual aspects of graphs prove to be a rich resource for this student by starting with the data of the transcript of a portion of the interview. Since most human exchanges are conducted on many levels and depend on much that is tacit and implicit, we augment this presentation of the "raw data" by a description of what we believe the issues between Dan and RN are at any time and how the dialog relates to them. At various stages throughout this presentation of the data, we also offer notes toward interpretations of it in clearly marked out sections. For the sake of comprehensibility, the portion of the interview to be presented has been divided into three distinct episodes, each of which has a coherence of its own and can serve as the basis for making certain points. At the end of each episode we analyze it in terms of our overall theme: the strengths, limitations, and potential for change of a fundamentally visual approach to graphs. There are five main sections to this paper. In addition to the present introductory section and three sections in which the three episodes are presented, there is a brief concluding section.

Episode 1

Overview. In this episode Dan works with straight upward heading Flow Rate graphs like the one in Figure 4. In predicting the shape of the Volume graphs and discussing relations between them, Dan focuses primarily on the steepness of graphs. However, when RN presses him to describe the difference between two Volume graphs on the screen that appear to an outsider to be distinguishable by one being straight while the other bends, Dan makes a contrast between them in terms of one being "wavy," while the other is straight.

Interview, Part 1. This phase of the interview begins with RN drawing the Flow Rate graph shown in Figure 4. He suggests that Dan sketch the Volume graph that would go with it.

Figure 4   Figure 5

Dan: The flow rate went up like that? That means that this thing [*the device*] will be all — I'd have all the air in here [*in the bellows*] and then I'd be pushing down [*on the bellows*]. So I think that the volume would go up from whatever that distance right there is [*points to the vertical intercept of the given Flow Rate graph*].

It would go up much more — rapidly, than say if we had this one
here — going up as that [*as he says this he runs his finger along an
imaginary line parallel to, but slightly lower than, the given Flow
Rate graph*]. Then it would be less [*draws the Volume graph, Figure
5*].

Dan is comparing the Volume graph he has drawn to one that he might have
drawn of a parallel Flow Rate graph that was lower than the given one. Picking
up on this note, RN asks him to relate what he has just done to an earlier case in
which Dan was given two Flow Rate graphs (shown in Figure 6) and eventually
drew the Volume graphs that are connected to them (shown in Figure 7). RN
then adds a new graph to the one shown in Figure 4 (shown as graph (2) in
Figure 8) and asks Dan for the Volume graph that would be connected to it.
Dan replies:

Figure 6   Figure 7

Dan: So let me think. When we did the flow rate that went like that
     [*referring to horizontal Flow Rate graph*], it went, just went up,
     like this. [*draws the Volume graph (2) in Figure 9*].

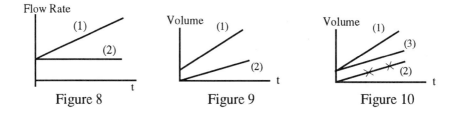

Figure 8   Figure 9   Figure 10

When RN asks Dan if he really wants his graph (2) in Figure 9 to start at 0,
Dan crosses it out and replaces it with graph (3) on Figure 10. RN then asks:

RN: Yes. And the difference is that the dark one —?

Dan: Would go more rapidly than the — light.

RN then suggests that Dan produce these Flow Rate graphs with the device in
order to see what happens.

Interpretive Notes.

- Dan frames his response to the two Flow Rate graphs given in Figure 8 in terms of one being more or less steep than the other. In drawing the Volume graph that goes with the given oblique straight-line Flow Rate graph, he draws a graph which is also an oblique straight line and differs from the given one only in being more steep. This visual attribute of steepness is a critical one for Dan in his anticipations of the relationships between Flow Rate and Volume graphs.

- In drawing Volume graphs that are conspicuously steeper than the given Flow Rate graphs Dan shows that he distinguishes between Flow Rate and Volume and uses these distinctions as they relate to the labels on vertical axes.

- Even though, Dan's responses to these graphs are framed in terms of visual attributes, he interprets them right from the start in terms of actions with the device. For instance, in the first exchange of this part of the interview he says: "The flow rate went up like that? That means that this thing [*the device*] will be all — I'd have all the air in here [*in the bellows*] and then I'd be pushing down [*on the bellows*]."

Interview, Part 2. Dan then works with the device for four minutes with the goal of producing Flow Rate graphs that look like those in Figure 8. In this time he explores the possible meanings of such distinctions as fast vs. slow and high vs. low in terms of the device and the kinesthetic actions to drive it. Through his experience in using the device, he must relate his own sense of terms such as these to the results of his physical actions and the readings on the screen. The graphs he produces in this process are shown in Figure 11. Then both RN and Dan look at the Volume graphs produced for these tries by the computer, which are shown in Figure 12. Dan's response to these Volume graphs is:

Dan: .... and, and, it, I think it, it didn't go as much drastically as I thought it was going to go. It was a little more closer than I thought it would be. It was a little more, yeah, it's closer than I thought it was going to be. I thought it was going to be more spread apart.

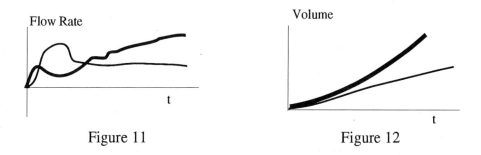

Flow Rate

Volume

t

t

Figure 11

Figure 12

In his response Dan focusses solely on whether or not the two Volume graphs are sufficiently *spread apart*. He does not remark on the shape of either graph.

RN then suggests that Dan try again, this time using the valve, which he does. Again, there is a period in which Dan works with the device in order to become better acquainted with the way in which it responds to his intentions and actions. The graphs that result from this attempt are shown in Figures 13 and 14.

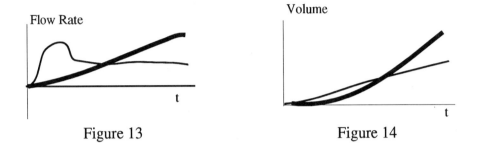

Figure 13   Figure 14

In response to these new Volume graphs he says:

Dan: Yes. I don't think it shows that much difference — between the constant flow rate and the flow rate that increased all the time. It didn't show very much difference in the volume.

And then, in response to RN's more pointed question of whether the dark Volume graph is a straight line, he says:

Dan: No, it's sort of like - eh. The other one is probably more straight than — the lighter one is probably straighter than the dark one.

In his first response to the graphs on the screen, Dan does not see the difference between them that he expected to see, that they be *spread apart*. In fact, because of the particular way in which he has used the device, the graphs are even less spread apart than they were the first time. However, in response to RN's direct question, he acknowledges that a distinction could be made.

RN then asks Dan if there is a reason for the dark graph being less straight than the light.

Dan: I don't know, the only reason could be that - flow rate increased, but it didn't increase that — so much per second that it - - increased more rapidly — not enough to make it a rigid line. [*As he says this he moves his hand along an imaginary straight oblique line in space.*] So it was just sort of like it increased, it increased [*gestures with his hands - making an undulating wavy line*], but it wasn't like very, very much of an increase the whole time throughout.

Dan's response is, then, that the dark graph is less straight than the light graph because, while the dark Flow Rate graph increased, its increase was not sufficiently great to make the dark Volume graph a straight line, "not enough to make it a rigid line." The result is that in the case of the dark Flow Rate graph the Volume increases in an undulating or *wavy* manner.

RN then asks how this description would be reflected graphically in his prediction of the dark Volume graph and Dan replies:

> Dan: .....Instead of being like that [*gestures toward the Volume graph (1) in Figure 10*]? I think, it probably should have been more — less — subtle [*as he says this, he draws the wavy graph (1) in Figure 15*] than it came out to be. Even — oh maybe not — because the amount of volume in there [*the light Flow Rate graph*] is always the same, so, that with the light one, even though the amount of air going in [*gestures with hand to make straight horizontal line*] was sort of more like a straight line [*draws a slightly straighter Volume graph (2) in Figure 15*], like a constant amount of air going in.......

In the first part of this statement Dan has given the graphical version of his previous statement about the dark Volume graph being wavy. He has drawn a wavy graph and asserted that it, and not the straight Volume graph (3) in Figure 10, is the Volume graph that should correspond to the given oblique straight Flow Rate graph. The term he uses for such a shape is that it be *subtle*.

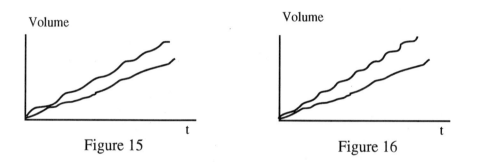

Figure 15   Figure 16

RN again presses Dan by asking for an explanation for the fact that the Volume graph (1) that he has drawn in Figure 15, is not as straight as the Volume graph (2) in that figure.

> Dan: Uh. [*Laughter*] Hmmm. [*Pause*] But even though maybe, even though the —[*draws a new set of axes*] the flow rate increased so much, but when we did it, it didn't increase so much — so maybe it was — it was sort of like, I'm going up [*speaks slowly as he draws the more wavy graph in Figure 16*] and like this. So it wasn't very much —

As Dan draws Figure 16, which is even more wavy than graph (1) in Figure 15, he does so with a conspicuous wavy gesture of the hand.

Affirming Dan's view, RN then says:

> RN: Oh, I see, like a wave [*makes an undulating gesture with his hand*].

> Dan: Yeah, like, it goes in [*also makes an undulating gesture with his hand*], and ... some of that air, [*not like*] rigid, like, whooo — all the air going.

Dan is illustrating, through his gestures, contrasting types of motion. The first
type is indicated by an undulating hand movement which corresponds to the
graphs he referred to as *subtle*. The other type of movement is indicated by a
direct and straight movement of his hand; it is accompanied by the sound *whooo*,
the kind of sound made by an object moving very rapidly. He uses the term *rigid*
to describe such motion. RN then confirms with Dan what it is about the flow
of air in this case that leads to this *subtle* or wavy graph. He says:

RN: Because it's not constant?

Dan: Yes. Constant amount of air going in every second, it's different.

RN: But there is always more and more and more.

Dan: Yes.

Interpretive Notes.
- Steepness is the focus of Dan's analysis in the first part of this passage, although the form it takes here is in terms of whether the graphs are *spread
  apart*.

- As Dan is pressed by RN about whether or not the dark graph is straight or
  not, he first acknowledges that a distinction can be made and then responds
  with a new distinction among visual attributes, that of a segment of a graph
  being straight or wavy.

- The way in which Dan frames things here suggests that the normally expected
  kind of flow is a *direct* or *rigid* one (that goes *whooo*), which corresponds
  to a horizontal Flow Rate graph and has a Volume graph which is an oblique
  straight line. This is in contrast to a flow which is undulating, has an oblique
  Flow Rate graph; and a Volume graph which is wavy or *subtle*. There is even
  a hint, in the way Dan frames these issues, of the second kind of flow being
  insufficient in some way ("it didn't increase so much.") and this is a source
  of dissonance that Dan senses in his interpretation. To him, a wavy or subtle
  flow, corresponding to an oblique Flow Rate graph is *less*, in some way ("it
  didn't increase so much") than a direct flow that corresponds to a horizontal
  Flow Rate graph. But this quality of the dark graph being *less* conflicts with
  the fact that the oblique (dark) Flow Rate graph is *higher* than the horizontal
  (light) Flow Rate graph and that the dark Volume graph is, in part, *higher*
  than the light Volume graph.

- As he has done from the beginning of the interview, Dan merges here in
  complex ways visual attributes with gestural expressions and physical actions
  on the device.

Analysis.
Throughout this episode, Dan's approach is couched in terms of the visual
attributes of the graphs. At first, given a straight Flow Rate graph, he draws
a Volume graph that resembles it in every way, except that it is *steeper* than
the Flow Rate graph. In supporting his answer, he begins to contrast the given
Flow Rate graph against one which is parallel to the given one, but is *higher*.

Later, he describes the Volume graphs he sees on the screen as less *spread apart* than he had expected. When pressed further by RN he describes one as wavy in contrast to another, which is straight.

Dan's use of the visual aspects of graphs is entirely different from the direct importation into the graph of the appearance of the underlying physical situation referred to in Iconic Interpretation. In fact, even while his approach is couched in terms of particular visual attributes, his use and interpretation of them is never direct or literal; he is not a captive of his visual responses. Throughout the episode he works to relate in a flexible manner the visual attributes with the events within the device, as well as his actions upon it. He says in the very first exchange in Part 1 "The flow rate went up like that? That means that this thing [*the device*] will be all — I'd have all the air in here [*in the bellows*] and then I'd be pushing down [*on the bellows*]." Later, while pointing to the horizontal and oblique Flow Rate graphs, he refers to them in terms of "the constant flow rate and the flow rate that increased all the time." In contrast to many situations in which students work with a graph on paper that represents a physical situation not present to the student, so that the student tends to work with visual features of the graph or qualities of physical events, Dan is working with a graph and a physical event that are both fully present to him and directly connected. Thus it is not surprising that he moves back and forth between them and works to see through the features of the graph to a variety of meanings and relations in the physical context.

Not only does Dan seek to ground these visual attributes in his experience of the device, but he also seeks to move in the direction of abstraction, of using these attributes as the terms, or tokens, of a system of reasoning about the situation. In his expectations of the appearance of the Volume graphs associated with given Flow Rate graphs, Dan implicitly uses rules or heuristic guides relating attributes of these graphs, such as "A Volume graph is steeper than its Flow Rate graph;" "The steeper the Flow Rate graph, the steeper the Volume graph." Such rules or guides can be seen as the beginnings of an attempt to construct a representational system around these attributes, with its own rules and operations, in order to more fully understand the situation.

Dan's focus on steepness as the significant feature of the graphs he is considering seems for a while to hinder him from seeing other aspects of the graphs. Having based his predictions about the Volume graphs on this visual attribute, he judges the Volume graphs on the computer screen in terms of it by first describing two graphs as not as "spread apart" as he expected, even though, to an experienced graph user, the more compelling distinction is that one graph is curved, while the other is not. In this way, Dan's understanding of the situation has some of the same rigid quality that is normally associated with the descriptions of students' misconceptions found in the literature [7, 2, 4]. However, it is not the case that Dan's approach can be characterized as stubbornly resistant to change, as is common among descriptions of student misconceptions, since, when he is pressed by RN about these graphs, he shifts his description of them to an entirely different polarity, the one between straight and wavy.

In their critique of the assumptions behind research on student misconceptions as well as the educational implications of this research, Smith, diSessa and Roschelle [11], propose an alternative to the idea so common within this literature that students' faulty conceptions must be *replaced* by more adequate conceptions. They suggest, instead that "The replacement of misconceptions [should give] way to knowledge refinement as a general description of conceptual change. Old ideas can combine (and recombine) in diverse ways with other old ideas and new ideas learned from instruction (p.47)." We propose that a related use of the term *refinement* is appropriate here to describe the ways in which Dan's understanding of this situation is changing. He has ideas about the graphs and about the physical events they refer to, which are to some extent faulty or inadequate. But they do not have the monolithic quality that would indicate that they must be replaced. Through his interactions with the device and with RN, in this episode and in those that follow, Dan continually modifies, extends, enriches and makes more complex his understanding of the situation. Learning for Dan is not the result of replacing what he knows, but of refining it.

Having presented one episode and its analysis, we wish to discuss briefly some issues of methodology of the interview and our interpretations of it. Dan's visual response to these graphs is a very rich resource for him. Clearly, such a response comes from within him and works very well for him. His commitment to his responses to the graphs and to the general approach they are a part of is immediately evident in the videotape of the interview and, we think, clear in the transcript. It could be remarked, however, that since Dan is asked questions about pairs of graphs that have familiar and notable shapes, his approach in this episode could be seen to some extent as a consequence of the form of the questions he is asked. This would present a difficulty for our analysis of this interview if we were claiming that Dan's responses reflected Dan's mental structures alone and that they were independent of the structure of the world around him. However, we do not make such claims. We believe that Dan, like all of us, is influenced by his various social and physical environments; we wish to analyze his responses within the environment of this interview. The question then is, what in this environment does Dan find salient, what does he make use of, and how does he use these environmental features? The device and the interview support a visual approach which is both a viable and productive one for Dan. Our goal is to describe the particular qualities of this approach, its strengths and limitations, the ways it changes and the ways it resists change.

A parallel issue about what Dan knows or understands arises in connection with RN's role in the interview. The question could be asked: Is RN conducting a clinical interview, in which case he might be expected to be far more neutral than he is, or is he teaching Dan, in which case a justification for his questions as teaching strategies might be expected? Our response again is that a strictly neutral stance on the part of an interviewer is appropriate in an investigation of student knowledge that is independent of the student's social environment, but that we do not claim to be describing such knowledge. On the other hand, for RN to be teaching would suggest that he has conceptual or mathematical goals for Dan, which he does not. RN's primary goal in this interview is to help Dan to articulate his evolving knowledge about the situation involving the graphs and the device. We do not see this knowledge as residing within Dan alone, but as developing in the interaction among Dan, RN, and the device.

Therefore, we view RN as a person in a conversation in which his goal is to learn as much as possible about what the other person thinks about the topic of the conversation. As is true of anyone in such a conversation, he must constantly make decisions as to when to press or probe and when to support the other person. In addition, there are times when it is appropriate to put forth his own view of the topic. These decisions are guided, though, by the goal of finding out what the other person knows and understands, even while acknowledging that this may be influenced by the conversation itself.

Episode 2

Overview. In this episode, Dan works with the Flow Rate graph in Figure 18 that goes up steeply from 0 very briefly and then goes down to 0 more slowly. At first, Dan interprets this graph as indicating that the Air Level should go up and then down. When he sees the Flow Rate graph that results from his attempt to produce such a pattern, he acknowledges that it is different from what he expects, but is unable to determine the source of this discrepancy. When RN shows him how to produce such a graph by always pushing up, but at different rates, Dan can immediately describe what RN has done, but can only slowly produce such an event with the device. Dan's prediction of a Volume graph that corresponds to the given Flow Rate graph reflects his view that the Flow Rate graph has two distinct events connected by a point of transition.

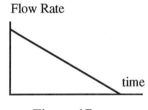

Flow Rate

time

Figure 17

Interview, Part 1. With Dan having closed the previous episode with a resolved view of the shapes of the Volume graphs that correspond to the given Flow Rate graphs, RN proposes a new Flow Rate graph. He asks Dan to sketch the Volume graph associated with a Flow Rate graph that has positive values but is headed downward (as in Figure 17). However, in Dan's response and the discussion that immediately follows, issues such as how to start with a non-zero Flow Rate begin to confound the discussion. In response, RN redefines the problem, changing the Flow Rate graph that Dan is to consider to the one shown in Figure 18.

The new Flow Rate graph has two segments, the first that heads upward and the second that heads downward. As Dan begins to sketch the Volume graph, in Figure 19, he says:

Dan: Well I, if it's like that [*Figure 18*] it had to — let me think. The
volume — The flow rate would go up, so that means the volume

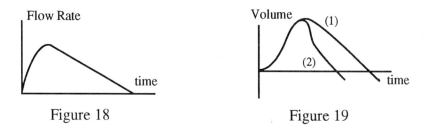

Figure 18   Figure 19

would also go up [*draws the first, upward, segment of the Volume graph in Figure 19*], and then it, it would drop. Let me think. [*Pauses, just when his pen reaches the top of the upward segment of the Volume graph*]. I'm trying to think if this [*the next segment he would draw*] would go very much below — if this [*the next segment*] would go below zero or not. Let me see. If the flow rate went up, volume increases, then it starts to decrease.

Dan first begins to relate the direction of the Flow Rate graph with the direction of the Volume graph. However, as he is about to draw the second segment of the Volume graph, another issue arises. He is "trying to think if [the second segment of the graph] would go very much below the horizontal axis." He goes on, pointing to the area where he expects to draw the second segment of the Volume graph, and says:

Dan: Well, the only thing I'm trying to think of is if this — as soon as I stop the increases, will it [*the Volume graph*] automatically just drop right down until it would be below here [*the horizontal axis*] and then it continues. [*As he says this, he outlines the shape of what he eventually draws as graph (2).*] Or will it just . . . It's either going to go like this [*draws graph (1)*]. Or it's either gonna go like this [*draws graph (2) in Figure 19*]. I'm not sure which one.

The Volume graph that Dan draws as belonging to the Flow Rate graph in Figure 18 has a segment that goes upward, but has two segments that go downward. He is uncertain which one he thinks is right. Both choices for the second segment go below the horizontal axis, but as Dan says, one (2) might "automatically just drop right down until it would be below here and then it continues," while the other one (1) does not fall so precipitously. However, as Dan deliberates on this issue, RN takes him back to the issue of the direction of the graphs by asking:

RN: It will go down, the volume? [Dan: Yes] It will go up a little and later will go down?

Dan: Go down. Well, I'm trying to — can't — I'm not sure if it will drop right away [*runs his finger along the first part of (2)*] and then come down [*runs his finger along the second part of (2)*] or would just gradually — —

RN: Or would go slowly? [Dan: Yes]

Interpretive Notes.

- Dan decomposes the graph into two separate segments, connected by a distinct transition point and treats the two segments as if they were graphs of separate events.

- He considers here for the first time the attributes of going up vs. going down, as well as being positive or negative.

- Dan expects that the direction of a segment of the Flow Rate graph will be literally reflected in the direction of the corresponding segment of the Volume graph.

- That Dan wonders about how fast the Volume graph will drop shows again that he consistently separates flow rate and volume as two related but distinct quantities in the situation.

Interview, Part 2. After making these predictions about the Volume graph, Dan begins to experiment with the device in order to produce a Flow Rate graph like the one he is given. He explores the device for a few minutes, rehearsing how he will push the air up and then down. When RN reminds him that the Flow Rate graph in Figure 18 describes flow rates that are always positive, he says:

Dan: Uh, I think you're right, I think it might have to always be above zero. I'm not sure though. If we go like this [*pushing the bellows down so that the Air Level goes up and watching the bag*]. No, it would go below zero, cause I pull the air in and then I'd be doing the reverse. So it should go below zero. Oh, I'll try it.

Although Dan uses the expression "pull the air in," one can see on the videotape that he *pushes* the air into the bag and then pulls it out. Not only does he act as if he believes that he must push and then perform an opposite action ("I'd be doing the reverse") but by first agreeing with RN's assertion, and then reversing what he has said ("No, it would go below zero."), he seems to acknowledge that this runs contrary to RN's assertion that the air flow is always positive. Notice that Dan refers primarily to the issue of whether or not his contemplated action (of pushing the air in and then pulling it out) would lead to the Flow Rate graph going below the horizontal axis; he does not question that this action is the one indicated by the Flow Rate graph given in Figure 18. RN does not challenge at this point Dan's apparent assumption that he must push and then pull, and Dan goes on with his experimentation.

Dan's first attempt at producing the given Flow Rate graph results in the graph in Figure 20 (The graph that appears on the monitor goes off the top of the screen because the window is not big enough. However, it is quite obvious how the missing part of the graph would appear. A dotted line has been drawn in the figure to indicate where the top of the screen is in relation to the graph.) When Dan and RN see it on the screen they both chuckle; Dan calls it a "fade," indicating that he sees that it is not at all like the given graph. RN points with his pencil to the part of the graph shown in Figure 20 by the arrow and suggests that this part of the graph is not correct. Dan replies:

Flow Rate

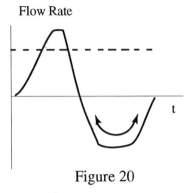

Figure 20

Dan: Well, I don't know. What am I supposed to do, push it down and
deal with this nice and slow and all? [*As he says this he is pointing
to the valve.*] [RN:] Oh.] Know what I'm trying to say? — how
much — it's going to be hard to get this [*points to the downward
segment of the given Flow Rate graph (Figure 18)*]. So I'm going
to have to pull it all the way out. [RN: Okay.] Or do you —
maybe I should just go, . . . maybe I should start like up here [*at
an intermediate Air Level*], go up and then go [*pull the Air Level
down*] nice and slowly like this. That's what I should do. Start it
say about, oh, 7. Start it about 7.

Dan agrees that the graph on the screen is not what they want, but expresses
frustration in trying to produce a better one ("What am I supposed to do....it's
going to be hard to get this."). He goes on, then, to wonder aloud about how the
given graph might be obtained. Perhaps if he would begin with the Air Level
very high ("maybe I should start like up here, go up") and then would pull the
air out sufficiently slowly ("go nice and slowly") then the graph on the screen
would not have the offending segment that goes below the axis. Dan feels that
he must keep going down "nice and slowly," but that he has been prevented from
doing so because he has reached the bottom of the bellows too soon. He expects
that starting higher will give him more room to keep moving down smoothly
before reaching the horizontal axis.

Dan then makes an attempt with the device and obtains a graph like that
shown in Figure 21. When he sees this graph he tries again and gets one like
that in Figure 22. He sees that the graph in Figure 22 does not match the given
Flow Rate graph, but, saying nothing, simply tries again. He gets another graph
which is almost identical to the one in Figure 22.

Interpretive Notes.
- Dan continues to respond to the Flow Rate graph under the assumption that
  the attributes of up and down indicate that the Air Level must go up and
  down.

- His first try, shown in Figure 20, at producing a Flow Rate graph like the one
  given in Figure 18 reflects Dan's focus on the attribute of the direction of the

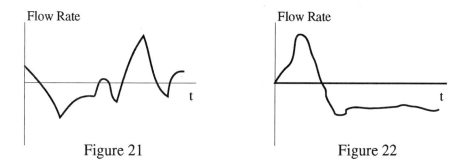

Figure 21   Figure 22

two segments in the given Flow Rate graph; he pushes the Air Level up for the first segment and pulls it down in the second segment, just as a literal reading of the graph would suggest. But then, when he observes the result on the screen (shown in Figure 20), he sees that his action of pulling the Air Level down, has led to a graph which not only goes down, but goes abruptly below the horizontal axis as well. This seems to suggest to him that there is something in the *manner* in which he pulls the Air Level down that would change the outcome. This is captured by his term "nice and slowly," which would seem to convey a quality of an action with the bellows that would avoid the sharp peak in the graph in Figure 20, followed by the precipitous plunge below the axis that he believes is caused by pulling too hard after the change in direction. After an attempt to obtain such a graph (shown in Figure 21) in which he seems primarily to be trying things out with the device, Dan makes two attempts that result in graphs like Figure 22. In these two attempts Dan has pushed the Air Level up and then pulled it down, but he has done the latter at a fairly constant rate. Dan has used the device in a way that could be seen as one enactment of his phrase *nice and slowly*.

Interview, Part 3. Pointing to the part of the graph in Figure 20 that goes below the horizontal axis, RN asks:

> RN: How could you do something that here goes down slowly without becoming negative?

With a thoughtful but dubious look on his face, Dan responds:

> Dan: Um, let me think. If I bring it — I have to bring it [*the Air Level*] down to zero maybe, go up, and then go down, down, down like that maybe [*pushes the bellows so the Air Level goes up a small amount and then slowly pulls the bellows, so the Air Level goes all the way down again*] — No, but, as soon as I pull down [*on the Air Level*], it [*the Flow Rate graph*] goes down to zero, so it has to — I can't. I don't think we can get that then.

RN then offers to try to produce the desired graph, to which Dan agrees. RN produces it once using the valve and once using the bellows. He obtains the graph in Figure 23. After Dan sees RN do this, he responds by saying:

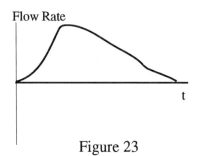

Flow Rate

t

Figure 23

Dan: Well — oh — It's just the amount of increase is less and less. I
see.

As Dan says this, he illustrates his understanding of what he has just seen by
the use of his forefinger and thumb. He holds them up, showing a gap between
them which he then makes smaller and smaller. Presumably, they are to indicate
"the amount of increase" that "is less and less." He then goes on to rehearse in
his mind how he will produce the same graph. He says:

Dan: Yeah, I see. So what do I do, just pull it down [*i.e. pull the Air
Level down and then let it up - as he says next*], let some air go up
and then we just let it go up slowly, is that what you did? Yeah.
[RN: Mm hm.] That's, that's different, I didn't think of that.

RN: Does it make sense?

Dan: Yeah, now it does.

RN: Let's try it.

Dan: It's tricky, though, tricky to figure that out.

RN: Why? What do you feel? [*Pause*] What is tricky . . .?

Dan: Oh — I thought that — as soon as I — see — I didn't think that to
get that line [*the second segment of the given Flow Rate graph*]. I
don't know why — I thought that — when, as soon as — to make
the line like that I had to pull it down. As soon as you pulled
down, it was below zero. I didn't think that maybe if we just did
this [*first pulls the Air Level down*] and then let the flow rate just
go nice and slow, that it would decrease from what it was before.
I should have thought of that, but I didn't.

Interpretive Notes.
• After producing his first Flow Rate graphs in Figures 21 and 22, Dan ponders
RN's question ("How could you do something that here goes down slowly
without becoming negative?"). As he does so, he considers the bellows and
the air bag, while he reviews what he believes he must do on the basis of the
given Flow Rate graph ("I have to bring it down to zero maybe, go up, and
then go down, down, down like that..."), as well as what he now sees as the

consequence of doing so ("no, but, as soon as I pull down, it goes down to zero..."). Dan believes he has been given a graph that could not possibly be obtained with the device. He is at an impasse ("I can't, I don't think we can get that then").

- As Dan watches RN producing the desired Flow Rate graph, his grasp of what RN has done seems almost immediate. He sees that the downward direction of the second segment does not indicate that the Air Level should go down, but that "the amount of increase is less and less." While saying this he holds up his fingers showing a gap between them which he makes smaller and smaller. Within the downward attribute of the second segment of the Flow Rate graph a new aspect emerges: the graph indicates that the amount of *increase* is less and less, in other words, the downward direction of the segment reflects simultaneously an increase *and* a decrease .

- Such a description is also a departure for Dan in that, for the first time, he has moved away from describing an entire interval in terms of a single attribute, such as up vs. down, and toward a description of a pattern of change in the flow of air where its quality changes with time during the time interval marked by the graph segment ("...that it would decrease *from what it was before*").

Interview, Part 4. RN then suggests that Dan try again to produce the desired Flow Rate graph, which Dan does. We see in Figure 24 the results of Dan's next six attempts to produce with the device a Flow Rate graph like the given one. In all these cases, the attempt is made with considerable deliberateness and care, with Dan verbalizing as he carries it out. As he makes each of these attempts, he watches the resulting Flow Rate graph develop, in real time, on the screen.

We can read something of what Dan is trying to do by the results he gets: Trial 1 shows him making a hard first push and then pushing at a constant rate. Trial 2 shows him making a hard first push, holding it steady, and then repeating this at a lesser intensity. Trial 3 is an exception in that he *pulls* the Air Level down part of the time. Trial 4 shows him again making a hard first push, and then a kind of wavering push, with Trial 5 a modification of this.

Interpretive Notes.

- In all these attempts, there is evidence in the graph of Dan briefly making a hard push down on the bellows at first and then going on to a different pattern of action. Thus, Dan continues his basic view that the event being described by the given Flow Rate graph has two segments separated by a distinct transition, one corresponding to the interval when the Flow Rate graph goes up and the other corresponding to the interval when the Flow Rate graph goes down. It is clear that Dan no longer thinks that the downwardness of the second segment of the Flow Rate graph indicates that the Air Level should go down, and that the issue he is working at is how to push the Air Level up in the appropriate manner. That is, he must clarify his notion of *nice and slowly* in such a way that the Flow Rate graph goes down, but does not plunge through the horizontal axis.

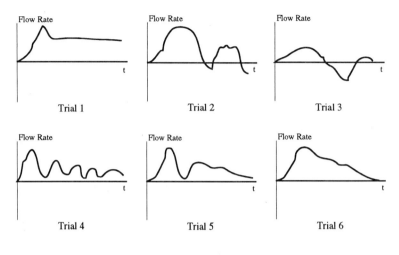

Figure 24

- By the fact that Dan can work with a situation in which the Air Level must be pushed up while there is a downward tendency in the positive Flow Rate graph, we see that he has succeeded in making an important first distinction between the two aspects of his efforts with the device, and, correspondingly, between the two aspects of the Flow Rate graph, whether the values are positive or negative and whether they increase or decrease. He is now aware that, on the one hand, there is the direction of the movement of the bellows and the Air Level, and, on the other hand, there is the manner or mode of this movement.

- Dan's learning experience in producing these graphs is a complex one. He says that he is to produce an event in which "the amount of increase is less and less," that he wants to "let the flow rate go nice and slow, that it would decrease from what it was before." It could be suggested that Dan has a completely clear idea of what he must do and is only having difficulty enacting it with the device. But, dealing with the device, Dan must interpret for the first time a feature of the graph he finds problematic: the transition from the upward to the downward segments. With his hand, Dan is expressing his intuition that the peak of the graph in Figure 18 indicates a discrete transition. Experimenting with the device Dan learns to moderate the transition kinesthetically. However even his final Trial 6 shows a distinct quality of abruptness in going from the fast increase to the slow increase. This issue, the nature of the peak of Flow Rate Graph 18, becomes the focus of the next segment.

Interview, Part 5. With Dan having succeeded in producing a Flow Rate graph that resembles the given one, RN asks him to predict the Volume graph that would go with it. Dan replies:

Volume

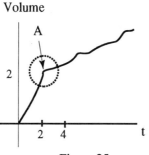

Figure 25

Dan: Always increasing — but — at the beginning it will increase. At
     the beginning [*draws Figure 25 on the drawing pad*] —. So —
     Volume — time. So, I believe that, the volume will go up, and
     then, and then it will just sort of like go up slower and slower and
     slower. I'm pretty sure that's what it will do.

Dan's description of the graph he draws is in terms of two segments, the first
in which "the volume will go up," and the second in which the volume "will just
sort of like go up slower and slower and slower." He is commenting on the overall
rates of change of the Volume in the periods represented by these two segments
and, presumably, the relative steepness of the two graphs.

RN then asks, pointing to the second segment of Dan's graph:

RN: This being a straight line?

Dan: It will — this will not be as — will not — it's incline [*indicates the
     entire first segment of his sketch of the Volume graph emphasizing
     its steepness*] will not be as much as — It will reach, it will reach
     this point here [*points to the place (A) on his sketch and at the
     same time points to Flow Rate graph on the screen*] and then it will
     start to even off sort of [*runs finger along entire second segment of
     his sketch*]. It might not, it might be a straight line, but I, I don't
     think it will. I think it will be just wavy, you know, it will be, it
     will start to.

In response to RN's question, Dan at first continues to comment on the relative
steepness of the two segments of the Volume graph, but then answers his question
directly in terms of the polarity he used earlier, of straight vs. wavy. However,
after making this one statement, he immediately returns to the issue of the
relative steepness of the two segments.

Dan: It will always increase, but it won't increase as — as the amount
     of time did here [*the first segment of his sketch*]. It won't increase.
     Say we did, say right here [*the first segment of his sketch*] — it
     was. I don't know, say it went up uh 2 liters [*beginning to write
     on the graph and points to first segment of his sketch*]. And we —
     very much less compared to this [*the first segment of his sketch*],
     and the following seconds after.

As he is speaking here, Dan makes the marks on the axes shown in Figure 25 and writes the numbers 2 and 4 near them. He then goes on to explain, in effect, that if the volume had gone up by 2 liters from the beginning of the graph to the time indicated by the point (A), and if the steepness of the second segment of the graph were equal to the steepness of the first, then one would expect the Volume to increase to 4 liters in the next time interval that is of the same length as the first one ("Right here, double whatever this is, it won't be 4 liters .... It won't, it won't be as much. The amount of air going in will not be as much as it did in this piece."). He indicates that this is not the case and his conclusion is that the second segment is less steep than the first.

Interpretive Notes.
- Here Dan shows more emphatically his inclination to think about the event as consisting of two distinct segments of time joined by a transition point. He has drawn two lines, one which is straight, and the other wavy.

- While Dan continues to use the term "slower" to describe what takes place in the second segment ("just sort of like go up slower and slower"), he also begins to use this word to compare what happens in this second segment to what happens in the first. This is in contrast to describing what happens within the second segment.

- Dan now uses two attributes in relation to the second segment of the graph that he used before — waviness and steepness. However, he now uses them simultaneously and with somewhat different meanings than they had before. He uses waviness to indicate that the air flow within this segment is slower and slower, while he uses steepness to compare the flow rate during this segment to the flow rate during the first segment.

Volume

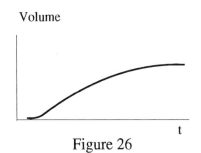

Figure 26

Interview, Part 6. In order to check Dan's prediction of the Volume graph, Dan and RN look at the screen together and see the graph shown in Figure 26. With a pained expression on his face, Dan sighs and says:

Dan: Nah. I don't know, it didn't really do that. Didn't have that amount of volume. It went up —....

When RN presses him about the discrepancy between his prediction and what he sees on the screen, Dan indicates with the mouse the graph he thinks *should*

appear on the screen, one with the same shape as his prediction of a Volume
graph shown in Figure 25. He goes on:

Dan: Maybe because that other line [*the first segment of the Flow Rate
graph*] was not the amount of — Um — flow rate was not so great .
. . [*to*] make it go up and then even off [*again traces the shape of his
own Volume vs. time graph*] like I thought it would. So, probably
had — not having the great amount of flow rate at the beginning
[*points to the beginning of the Volume graph on the screen, where
it is horizontal*], it just sort of went up, and up, and up [*pointing
to the middle section of the Volume graph on the screen*], but it —
it — I think it, basically it did what I thought it was going to do.
That it just sort of evened off here [*the end of the graph on the
screen*].

Interpretive Notes.
- Dan knows what he thinks the graph should look like, but is perplexed by
what he sees on the screen. Faced with the Volume graph on the screen that
lacks the key features of two segments which are joined at a point and which
are of markedly different steepness, Dan focuses on a single discrepant feature
between his prediction and the graph on the screen: that the graph on the
screen lacks the sharply upward segment at the beginning. He ascribes this
discrepancy to his idea of the event he has produced as "not having the great
amount of flow rate at the beginning." He attempts to say that the difference
is not a significant one by saying "I think it, basically it did what I thought it
was going to do." That is, the graph they see on the screen really does follow
the pattern of the graph he has drawn: "it just sort of evened off here."

- Just as was the case in Episode 1, this episode ends with Dan having a firm
idea of what the Volume graph of the event should look like, but having an
unresolved sense that what he sees on the screen does not contradict this
expectation, but also does not quite confirm it.

Analysis.
While there is in this episode an extension of Dan's visual approach to the
shapes of graphs, we see clearer manifestations of both persistence and change
within it. We also see, more sharply than earlier, how Dan's understanding of
the situation evolves because of the rich connections he makes between his visual
responses to graphs and other modalities of his experience such as his kinesthetic
sense of the bellows and the flow of air.

The Flow Rate graph that Dan deals with in this episode is more complex than
the one he dealt with in Episode 1, and his work with the attributes of graphs has
grown more complex along with it. Because this Flow Rate graph has segments
that go up and down, these attributes become significant in his responses for the
first time. He initially assumes that this means that the Volume graph must go
up and down and thinks it might also be negative, so that the polarity between
the graphical attributes of being positive or negative arises for the first time.
In addition, this raises new possibilities for the ways in which attributes of the
Flow Rate and Volume graphs could be connected. In Parts 5 and 6, when, as

a result of RN's intervention, Dan has come to realize that a downward heading Flow Rate graph does not indicate that the Air Level goes down, he connects the direction of the Flow Rate graph with the manner or intensity of the push on the bellows and with the relative steepness of the Volume graph. A Flow Rate graph that has the pattern of up-and-then-down is associated with a Volume graph that is more-and-then-less-steep. We see then that these visual attributes are not just labels Dan places on the graph segments as part of a static image of the situation. They are the terms of an expanding analysis of it that he revises in response in his interaction with RN and the device.

Dan reaches an impasse in this episode which he is then able to resolve, not only because of RN's intervention, but because the meanings he attaches to these attributes open out to other modalities of his experience. At the beginning of the episode, in Parts 1 and 2, we see Dan acting on his strongly held, but unexamined, assumption that when the Flow rate graph goes up or down, the Air Level will also go up or down. When he first acknowledges in Part 2 that the graph he has produced on the screen is different from the graph he predicted, he explains the difference between these graphs in terms of the *manner* in which he has pulled the Air Level down. This is a departure for Dan, because it is about a quality of an action, something that happens during the time interval marked by a segment of a graph. But still Dan cannot uncover the source of the contradiction, so that, in Part 3, RN is led to show him how such a graph can be produced. Having seen RN produce this event and having verbalized on what RN has done, Dan understands at some level that the direction of the Flow Rate graph does not determine the direction of the movement of the Air Level or the direction of the Volume graph. Nonetheless, Dan must still discover how to play out kinesthetically his incipient new ideas on the meaning of direction in a Flow Rate graph. Through his use of the device, with its real-time feedback, guiding his arm and his hand, watching the results on the screen, and verbalizing on what he is doing, Dan arrives at the view that the direction of the Flow Rate graph reflects the intensity with which the bellows is pushed. Dan's learning in this episode does not just consist, though, of the revision of one of his heuristic rules for relating the shape of the Flow Rate and Volume graphs. Through this experience, Dan has clarified the ideas he has about the behavior of the flow of air and how this relates to the information in a graph. In particular, he must work out the meaning of a flow rate that goes "nice and slowly," as well as how a transition from one graph segment to another is realized in the flow of air and his use of the device.

We see in this episode a wide array of changes and persistences in Dan's understanding of the situation. He begins the episode with the view that the direction of the Flow Rate graph indicates the direction of the movement of the Air Level and ends with the view that the downward direction of the Flow Rate graph indicates, instead, that the Flow Rate decreases. Throughout this time, he never varies from his view that the Flow Rate graph consists of two segments that describe two separate events with a distinguishable transition point between them. At the beginning of the episode he responds to the Flow Rate graph primarily in terms of its visual attributes, and at the end he again does so, but in a much more complex way. In the middle of the episode, he focuses almost exclusively on issues of quality of flow and how to produce certain effects with the device. Even more than was the case in Episode 1, Dan does not seem to be in a process of overcoming a misunderstanding, or of *replacing* one understanding of this situation with another. As was true then, Dan's learning is better described in terms of a process of *refining* his initial understanding into one that is somewhat more adequate.

When we consider the many changes and persistences in Dan's understanding of the situation that have taken place in these two episodes, we see *why* a process of refinement of a student's understanding of such situations fits so well in the present interview and is likely to be the process found in most learning situations. If knowledge of a situation is tightly bound together into a monolithic and coherent whole, then it can only change all at once, it can only be replaced. However, as Smith, diSessa, and Roschelle [11] propose, knowledge of a particular situation is not generally a unitary and monolithic whole; they "argue for an analytical shift from single units of knowledge to systems of knowledge with numerous elements and complex substructures that may gradually change, in bits and pieces and in different ways." (p.48). Such a description fits Dan's knowledge very well. Clearly, it is made up of many interlocking parts: of an overall view of the graph as made up of two segments, of segments having different meanings from different sources, of attributes of segments of the two graphs being associated in different ways, and of a kinesthetic sense of air flow. As these authors would put it, this system of knowledge is likely to change slowly "in bits and pieces." Refinement, not replacement, of this system is likely to be the more adequate description of the process of change.

We also see more in this episode how, rather than being just a source of errors and confusion, Dan's response to the visual aspects of these graphs is a rich resource and a productive basis for his learning about the situation. As we observed in Episode 1, even though Dan responds so consistently to the visual attributes of the graphs, his overall approach is in terms of qualities of air flow and his own actions on the device that can give meaning to these attributes. At the moment in this episode when his response to the shape of graphs seems most driven by pictorial qualities, when Dan acts so automatically on his implicit belief that when the Flow Rate graph is up or down, the Air Level must go up or down, he stops attending solely to visual attributes of whole segments and begins to bring his kinesthetic sense of air flow into play. Through his interaction with RN and the device, he is then not only able to move beyond this overly literal response to the shape of the graph, but he also modifies the meaning he finds in the direction of the graph. As has been the case in much of the interview, Dan not only sees the graph as a visual object, but, in Mason's terms, he sees *through* the graph to an increasingly wider field of meanings and relations beyond it.

Episode 3

Overview. RN helps Dan choose the proper shape for the second segment of the Volume graph by pointing to the place where the Flow Rate graph crosses the horizontal axis.

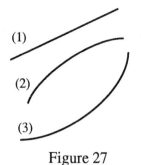

Figure 27

Interview. In order to pursue further the issue of the shape of the Volume graph they have been working on, RN presents Dan with the three graphs shown in Figure 27 and asks him which graph among these he thinks is most likely to be the Volume graph that corresponds to the second segment of the Flow Rate graph shown in Trial 6 of Figure 24. Dan's reply is:

> Dan: Which one of these is happening in the volume? [RN: Yes.] This, the first one [*points to the straight line (1) in Figure 27*] ... If this thing [*graph (2)*]] kept on going and going?
>
> RN: Yeah. I mean, just looking —
>
> Dan: I think you ended up [*points again to the straight line (1) in Figure 27*]. No, it has to always go up. It can never go down [*points to (2) in Figure 27*].

RN then flips back to the page on their drawing pad where the graph in Figure 18 is shown, and, pointing to the place where the graph touches the horizontal axis, says:

> RN: Now, when flow rate touches zero. [Dan: Mm hm.] Then what happens with volume?
>
> Dan: The flow rate touches zero? That means there's no air going in? [RN: Right.] Then that, it just sort of, I don't know, it, it would probably do it like number two then [*in Figure 27*]. [RN: This?] Yes. Go up and then sort of goes down.

RN is pointing to the place on the given Flow Rate graph where the graph touches the horizontal axis, which Dan immediately recognizes as a place where "there's no air going in." This leads Dan to quickly change his mind, and to say that the correct Volume graph would be the graph (2), which goes "up and then sort of goes down." They then continue.

> RN: But you are not sure? Just looking at that, it may be any of these three?
>
> Dan: Well, ... I think, it probably, it would be this [*points at the graph (2) in Figure 27*].

RN: This one? [*He points at (2) on Figure 27.*] [Dan: Yes.] It couldn't be this one [*(3) on Figure 27*]? Could be? [Dan: No.] Why?

Dan: Because the — in this one [*(3)*] the — the volume is always increasing [*running his finger along (3)*] and we said the flow, it [*the Flow Rate*] went down to zero so it [*the Volume*] had to, it had to, it had to stop and go back down.

RN: This [*(2) on Figure 27*] is the only one that stops? [Dan: Yes.] Okay.

Interpretive Notes.
- In his first reply to RN's question, Dan asserts that graph (1) must be the correct graph. He knows that the volume goes up ("No, it has to always go up."), and he regards graph (2) as going down. However, since the direction of the graph (2) is up, we see that the attribute of "down-ness" has now come to have a different meaning from its earlier one. The sense in which this graph goes down in Dan's mind is indicated by his question: "If this thing [*graph (2)*] kept on going and going?" That is, this graph goes down because a *potentiality* exists for its going down—if the graph were extended.

- RN's suggestion, that Dan focus on the single point where the Flow Rate graph crosses the horizontal axis, enables Dan to come to a resolved answer, that graph (2) is the correct choice. His reason is that at such a point "there's no air going in," so that "it [*the Flow Rate*] went down to zero so it [*the Volume*] had to...to stop and go back down." While he no longer explicitly mentions graph (1), he is now able to reject graph (3) because "in this one [*(3)*] the — the volume is always increasing [*running his finger along (3)*]." All three of these graphs go up; they differ only in curvature. Dan is now making distinctions among them based on his interpretations of curvature in terms of tendencies of the flow of air toward a certain state.

- RN's question, "Now, when flow rate touches zero. Then what happens with volume?" is about both the graph ("flow rate touches zero") and the flow of air ("what happens with the volume?"). But this presents Dan with no problem; he immediately responds in terms of the flow of air ("That means there's no air going in?"). Such a pattern can be seen in other exchanges in this episode. As is common in the conversation between graph users, the distinction between a representation and the phenomenon it represents is barely made.

Analysis.
This episode is a coda for our analysis of the interview in that it demonstrates again the two main points we wish to make about Dan's learning. First, we see an even easier movement on Dan's part between the graph as a visual object and the physical events and relations within the device. The graph is increasingly *transparent* to Dan.

We also see again the inherent openness and flexibility of Dan's approach to the use and interpretation of graphs. Asked to choose from among the three graphs RN proposes, Dan initially makes his choice according to an as yet unresolved sense he has about the attributes of up and down. He knows that the

volume goes up, but there is something about graph (2) that goes down. As soon
as RN suggests that he consider the single point at which the Flow Rate is 0, the
quality of "down-ness" of a graph is sharpened by Dan to relate to a tendency
in the flow rate to go toward 0. As was the case in the earlier episodes, this
change in Dan's understanding of the situation, does not consist of substituting
or *replacing* one idea for another, but of supplementing his knowledge system
with a new idea and connecting it to other ideas within the system. There is no
inherent contradiction between interpreting whole segments of graphs by their
visual attributes and interpreting segments on the basis of the behavior of the
air flow at a single isolated point. Dan is *refining* his understanding of this situation in that he has enlarged it and made more connections among its various
"pieces."

Conclusion

We have documented and analyzed the interactions between Dan, RN, and
the Air Flow device in order to give a full sense of Dan's fundamentally visual
approach to constructing and interpreting graphs, one in which he proceeds from
a response to the visual attributes of graphs as he moves outward to a more
complex understanding of the functional situation. We have tried to show that,
while Dan's understanding of this situation is, in ways, inadequate, unresolved
and incomplete, it has undergone considerable change within the period of this
interview and could be expected to change even more in subsequent experiences
of this kind. We have provided such a detailed record of this interview, in part,
because we believe that an elucidation of Dan's thinking and the subtle changes
it undergoes, in their fullest complexity, will demonstrate that the process of
change in Dan's understanding of this situation is one of gradual refinement of his
ideas and not one of replacement. Indeed, we believe that an image of change by
replacement, which is a natural response to the great attention given by several
researchers [2, 7] to student errors in graph reading, is mistaken because these
errors are in many cases artifacts of studying students' understanding through
written tests or narrowly focussed interviews in which they must choose between
a small number of possible answers, rather than exploring ideas and shifting
points of view. This session with Dan is typical of interviews in more than
fifteen teaching experiments we have conducted with high school and college
students, using a variety of physical devices, in that Dan's visual response to
graphs is a productive one and not merely a source of errors and confusion.

In the standard approach to teaching graphing, one starts by drawing coordinate axes, which are carefully labeled and scaled, so that every point on the plane
has an "address" in relation to these axes—a pair of numbers called its coordinates. Graphs are then interpreted through the implicit, but deceptively simple,
rule: *A graph is made up of its points.* These points are immediately associated
with their coordinate pairs, so that the functional relationship described or represented by the graph consists of the correspondences "x_0 corresponds to y_0"
for exactly those coordinate pairs (x_0, y_0) associated with points on the graph.
But in the same sense that playing chess is not just a matter of knowing the
rules of the game, graphing involves much more than associating points with
their coordinates. Expertise in graphing is reflected by the ability to recognize

the meaning of visual attributes, to extract information from critical points, to make inferences about how change in one variable leads change in the other, to preserve functional relationships under changes of scale, to know when the shape of a graph is significant or insignificant, and to "see through" the graph to the represented situation. The knowledge about a functional relationship one gathers through such processes is *mathematically* justified by reducing the properties of a graph to its behavior at a point, but this does not mean that skill in carrying out these processes develops solely from an ability to interpret the meaning of a graph from the information in its points. To assert that a graph increases, but that it is less and less steep, one must make an argument that *ultimately* reduces to the values of the graph at individual points, but one might believe that this statement is true for other reasons having to do with more global properties of the graph. Moreover, expert use of such knowledge can never be reduced to a set of explicit rules, in part, because their use is contingent upon the features of the particular functional situation at hand. If a problem involves rate of change, then it is highly significant if an upward heading graph is less and less steep, but if the problem involves accumulations, this property has relatively little significance. Consequently, developing such expertise will always involve the use of actual graphs in a diversity of contexts and purposes.

Our analysis suggests an alternative approach to teaching graphing based on activities in which students explore graphing situations using *all* they know about graphs, playing with visual attributes, making predictions, and communicating about relationships within curves. Such approaches, for students at a variety of grade levels, are being developed by a number of curriculum designers, including, among others, Barnes [1], Hughes-Hallet & Gleason [3], Swan [12], Yarushalmy & Shternberg [17], and Tierney, et al [15]. We believe that the nature of Dan's learning in this interview highlights the potential value of such curricular projects. Dan comes to this interview with a background of ideas and experiences about graphing gathered from instruction in mathematics and other school subjects, as well as television and newspapers. What he learns in this interview is not the result of appealing to the rules governing the relationship between points and coordinates, but of experimenting with the different meanings he comes to recognize in the visual attributes of a graph. From the very beginning Dan exercises his ability to "see through" actions on the device corresponding to specific visual attributes. When he finds it valuable to do so, he makes use of the information conveyed by specific points and their coordinates (e.g. his comparison in Episode 2 between intervals from 0 to 2 seconds and from 2 to 4 seconds, and his focus in Episode 3 on the time at which Flow Rate is zero). Rather than replacing his ideas with new ones, Dan begins to develop expertise by enriching the diversity and connectedness of meanings he constructs, starting with visual attributes of the graph.

That Dan's visual approach is so productive for him is partly the result, we believe, of the learning environment he is working in. Because of the direct link between the graphs on the computer screen and the events in the Air Flow device, Dan can interpret the visual qualities of the graphs directly in terms of his own actions and the behavior of the flow of air. This can be seen in the first line of the transcript where he says, looking at the graph: "The flow rate went up like that? That means this thing [*the device*] will be all — I'd have all the air in here [*in the bellows*] and then I'd be pushing down [*on the bellows*]." This link between visual attributes and situated actions is emerging as a critical element in students' learning. Nemirovsky [**9**], Thornton [**13**], and Tierney et al [**15**] have investigated how activities with a motion detector — in which the movement of objects or the student's body produces, in real time, graphs of distance or velocity vs. time — can produce significant differences in students' understanding of motion graphs. Framed by these themes, the chief contribution we hope to make with this paper is a deeper understanding of the possible routes by which a student could come to have expertise in making and interpreting graphs.

REFERENCES

1. Barnes, M., *Investigating Change. An Introduction to Calculus for Australian Schools*, Curriculum Corporation., Carlton South, Australia, 1991.
2. Clement, J., *The concept of variation and misconceptions in Cartesian graphing.*, Focus on Learning Problems in Mathematics **11(1-2)** (1989), 77–87.
3. Hughes-Hallet, D. & Gleason, A., *Calculus. Preliminary Edition.*, John Wiley and Sons, Inc., New York:, 1993.
4. Janvier, C., *Translation processes in mathematics education. In C. Janvier (Ed.)*, Problems of representations in mathematics learning and problem solving, Lawrence Erlbaum Associates, Hillsdale, NJ:, 1987, pp. 27–31.
5. Leinhardt, G., Zaslovsky, O., & Stein, M., *Functions, graphs, and graphing: Tasks, learning, and teaching.*, Review of Educational Research **60(1)** (1990), 1–64.
6. Mason, John H., *What do symbols represent?*, C. Janvier (Ed), Problems of Representation in the Teaching and Learning of Mathematics, Lawrence Erlbaum Associates, Hillsdale, N.J.:, 1987, pp. 73-81.
7. McDermott, L., Rosenquist, M. & vanZee, E., *Student difficulties in connecting graphs and physics: Examples from kinematics.*, American Journal of Physics **55(6)** (1987), 503-513.
8. Monk, G.S., *Students' understanding of functions in calculus courses. Unpublished manuscript* (1987), University of Washington, Seattle.
9. Nemirovsky, R., *On ways of symbolizing: The case of Laura and velocity sign.*, The Journal of Mathematical Behavior (in press).
10. Nemirovsky, R. & Rubin, A., *Students' tendency to assume resemblances between a function and its derivative*, TERC Working Papers (1992).
11. Smith, J., diSessa, A., & Roschelle, J., *Misconceptions reconceived.*, Journal of the Learning Sciences (In press).
12. Swan, M.(Editor), *The Language of Functions and Graphs*, Shell Centre for Mathematics Education, University of Nottingham, 1985.
13. Thornton, R & Sokolov, D., *Learning motion Concepts using real-time micro-computer-based laboratory tools*, American Journal of Physics **58(9)** (1990), 858–866.
14. Tierney, C., Nemirovsky, R., Weinberg, A. (In press), *Changes: Up and Down the Number Line. Curricular unit for grade 3*, Dale Seymour Publications., Palo Alto, California:.
15. Tierney, C.; Nemirovsky, R.; Wright, T.; Ackermann, E., *Body motion and children's understanding of graphs.*, J. Rossi & B. Pence (Eds.) Proceedings of the XVI Annual Meeting of the Psychology of Mathematics Education- North American Chapter, vol. 1, The Center for Mathematics and Computer Science Education, San Jose State University., San Jose, California:, 1993, pp. 192–198.
16. Tierney, C. Weinberg, A. Nemirovsky, R. (In press), *Graphs: Changes Over Time. Curricular unit for grade 4.*, Dale Seymour Publications, Palo Alto, California.
17. Yarushalmy, M. & Shternberg, B., *The Algebra Sketchbook*, Sunburst, Pleasantville, NY, 1991.

UNIVERSITY OF WASHINGTON AND TECHNICAL EDUACATION RESEARCH CENTER

CBMS Issues in Mathematics Education
Volume 4, 1994

The Effect of the Graphing Calculator on Female Students' Spatial Visualization Skills and Level-of Understanding in Elementary Graphing and Algebra Concepts

MARY MARGARET SHOAF-GRUBBS

ABSTRACT. The purpose of this study was to explore the role of the graphing calculator in the enhancement of spatial visualization skills and mathematical understanding for female students entering college with weak abilities in mathematics. Nineteen students in one section of elementary college algebra used the graphing calculator. Eighteen were members of the traditional group in which the material was presented in as nearly identically a fashion as possible, but without the graphing calculator. Gains for nineteen tests (of which four were aggregates of the other fifteen) covering spatial visualization and level-of-understanding in elementary algebra and graphing concepts were examined. The gains shown by the Calculator Group showed t-test significance at the $p = .05$ or less level in ten of the nineteen test situations examined.

INTRODUCTION

The study reported in this paper examines the role that technology can play within two areas of mathematics learning, spatial visualization and mathematical understanding. McGee [23] defined spatial visualization as those tasks that "involve the ability to mentally manipulate, rotate, twist, or invert a pictorially presented stimulus object." Kersh and Cook [18] extended this definition to include either the rotation or transformation of a mental object. This study uses Kersh and Cook's extended definition.

In a factor analytic studey Conner and Serbin [6] reported that spatial visualization skills appear to contribute meaningfully in predicting mathematics achievement. They found that the Paper Folding Test and the Card Rotation

This article is based on the author's dissertation at Columbia University under the direction of Dr. Henry O. Pollak. Dr. Shoaf-Grubbs wishes to thank Dr. Alan Schoenfeld, Dr. Cathy Kessel, Dr. Judith T. Sowder, and one anonymous reviewer for many helpful suggestions.

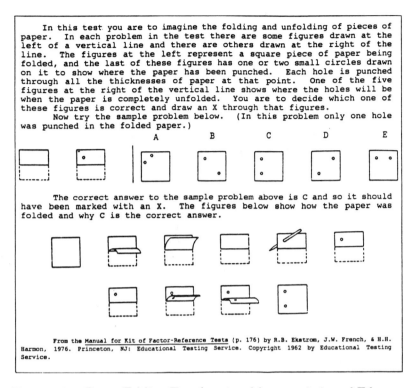

In this test you are to imagine the folding and unfolding of pieces of paper. In each problem in the test there are some figures drawn at the left of a vertical line and there are others drawn at the right of the line. The figures at the left represent a square piece of paper being folded, and the last of these figures has one or two small circles drawn on it to show where the paper has been punched. Each hole is punched through all the thicknesses of paper at that point. One of the five figures at the right of the vertical line shows where the holes will be when the paper is completely unfolded. You are to decide which one of these figures is correct and draw an X through that figures.

Now try the sample problem below. (In this problem only one hole was punched in the folded paper.)

The correct answer to the sample problem above is C and so it should have been marked with an X. The figures below show how the paper was folded and why C is the correct answer.

From the Manual for Kit of Factor-Reference Tests (p. 176) by R.B. Ekstrom, J.W. French, & H.H. Harmon, 1976. Princeton, NJ: Educational Testing Service. Copyright 1962 by Educational Testing Service.

FIGURE 1. Paper Folding Test (reprinted by permission of Educational Testing Service, the copyright owner).

Test [9] measure spatial visualization. The Paper Folding Test (see Figure 1 for sample items) consists of two three-minute tests with ten items on each test. The student must select the drawing that shows where the punched hole would appear on the paper when it is fully opened.

The Card Rotation Test (see Figure 2 for sample items) also consists of two three-minute tests numbering ten items. The student is to indicate whether the shape is the same as the original or if it has been turned over. These two tests measuring general spatial visualization ability were chosen for this study because it is necessary for the student to mentally move, rotate, or alter all or part of the representations of the functions studied in the course.

Researchers have tried to identify factors related to mathematics achievement and understanding in efforts to determine why some students learn and understand mathematics better than others. Because studies have reported positive correlations between spatial skills and mathematics performance [31] spatial skills are of special interest in mathematics education and the mastery of mathematical concepts. Tartre [31] has stated:

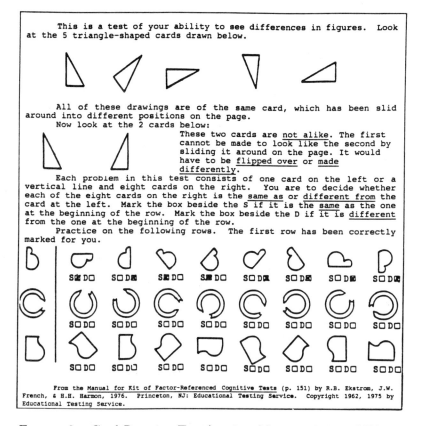

FIGURE 2.   Card Rotation Test (reprinted by permission of Educational Testing Service, the copyright owner).

spatial skill may be a more general indicator of a particular way of organizing thought in which new information is linked to previous knowledge structures to help make sense of the new material (p. 227).

Tartre [32] further describes a three-year study in which male and female subjects were classified as either high or low in both spatial visualization and verbal skills. Females consistently obtained the lowest scores on the mathematics tests regardless of their classification in either spatial or verbal ability when compared with their male peers. When the groups were offered help in translating verbal information into a pictorial representation, females with low spatial scores required the most help, while females with high spatial scores required the least help. Furthermore, females with high verbal but low spatial skills fell further behind the other groups over the 3 year study.

By 1986 research results were reporting smaller but consistent gender differences favoring males in the cognitive domain of spatial visualization [35]. Although more recent meta-analyses no longer find significant gender differences in spatial visualization, [32], spatial visualization skill remains an important predictive indicator for females within the realm of mathematics. Fennema [10] has stated that spatial visualization is logically related to mathematics. Sherman [29] found spatial visualization to be a discriminator more important for girls than boys in predicting the number of mathematics college preparation courses taken in high school. Fennema and Tartre [11] suggest that low spatial visualization skills may be more debilitating to girls' mathematical problem-solving skills than to boys'. Thus spatial visualization skills could serve as a means, perhaps more often in the case of females, to identify the students weak in the skills acknowledged to predict success in mathematics [31].

Recent investigations support the hypothesis that gender differences in spatial visualization exist and provide evidence that these skills can be learned, improved, and enhanced through training [2, 1].

Cartledge [4] recommended improving females' spatial visualization skill as one way to overcome the superiority (if, indeed, it exists) of males in mathematics achievement from grades eight through college.[1] A correlation exists between students' participation in spatial activities and their measured spatial ability, with males participating in a greater number of spatial activities [26].

Studies in computer technology and mathematics education have confirmed the positive effect of calculators and computers in the classroom [5]. Because gender differences in spatial visualization are considered by some to be a reason for gender differences in mathematical achievement, research indicating improved spatial visualization with calculator use is particularly encouraging. Vazquez [37] reported gains in spatial visualization skills for students using fraphing calculators. Dunham [8] found that pre-test differences favoring males on visual graphing items were not evident on the post-test after instruction with graphing calculators.

Hyde, Fennema, and Lamon [16] performed a meta-analysis of 100 studies reporting gender differences in mathematics performance and found that "even among college students or college-bound students, gender differences are at most moderate" (pg. 151). However, they recommended caution in interpreting these results, for when differences are found they exist in critical areas and require attention. Because success and skill in problem-solving are critical in many mathematics-related fields, it is likely that gender differences could remain a "critical filter" for females and that particular concern should be focused on gender equality in the area of mathematics education.

Mathematics can be learned by all types of students. But, a method of teaching or presenting a mathematical concept so that one person can successfully understand and use the mathematics may not work for another concept or another person. Attention has focused on how "mathematics teaching and learning can be improved by developing more powerful approaches to connecting thinking

---

[1]Earlier studies have frequently reported wide gender gaps in mathematics achievement. The recent meta-analysis by Hyde, Fennema, and Lamon [16] shows that according to more recent studies these gaps have been shrinking. See comments to follow.

and mathematics" [30]. New concepts should be taught as extensions of prior mathematics understanding, thereby enabling students to see connections from one concept to another and develop a stronger conceptual understanding during the learning process. Students receive higher scores on a mathematics achievement test when teaching methods include concrete, manipulative materials [24]. Horowitz [15] supported this deduction and reported that the visualizability of a problem affects its solvability by lower performance subjects.

Technology might be a mechanism through which these goals could be realized. The Calculator and Computer Pre-Calculus Project (C^2PC) [7] indicates that mathematics instruction with graphing technology can have a positive impact on student achievement and mathematical understanding. Students using graphing utilities attain higher levels of understanding in graphing concepts than those in a traditionally taught mathematics classroom [33, 12]. Ruthven [27] reported research conclusions in the area of gender illustrating that:

> access to information technology can have an important influence both on the mathematical approaches employed by students and on their mathematical attainment. On the symbolisation items, use of graphic calculators was associated not only with markedly superior attainment by all students, but with greatly enhanced relative attainment on the part of female students. (p. 431)

By converting an equation to a visual mode graphing calculators allow students to explore and to make connections between mathematical concepts [36, 40].

This study examines spatial visualization and understanding in mathematics and the role technology plays in their enhancement.

METHOD

Subjects.

The subjects for the study were 37 students enrolled in two elementary college algebra classes at an all-women's liberal arts college. Prior to registration the researcher and her committee randomly labeled the two course sections as Calculator and Traditional. The students randomly enrolled in one of the two class sections after scoring below the pre-calculus level on the college's in-house mathematics placement test. The section chosen by each student was determined by her other scheduled courses for the semester. Students were not aware at the time off registration that they would be participating in a research study. It was only after meeting the class that they were made aware of the research study or which section would be using the graphing calculator. Nineteen were in the Calculator/Experimental Group using the graphing calculator, and 18 were members of the Traditional/Control Group in which the material was presented in as nearly identically a fashion as possible, but without the graphing calculator. No attrition occurred in either class section and all students took all tests. The traditional group used paper-and-pencial and overhead transparencies to draw and study graphs of identical functions whereas the calculator group used the graphing calculator.

Materials and Procedure.

The Keedy/Bittinger text, *College Algebra – A Functions Approach*, fourth edition, published by Addison-Wesley was used by both groups. The instructor, lesson plans, and number of graphs examined during the class period were the same. Any discovery by a student in one group that had not been included in the lesson plans was prepared and presented to the other group. Sketches drawn by hand or those rendered by the graphing calculator enabled the instructor to create a highly visual exploration-of-concepts teaching and learning environment. All three methods of concept presentation and problem solution – numerical, algebraic, and graphical – were part of the lessons. Students actively engaged in class discussion by stating, re-defining, and/or modifying conjectures about the concepts being examined. This then aided them in drawing conclusions that would support a conjecture. Class dialogue helped the instructor to determine areas where some students encountered difficulty such that more time and explanation could be given to those topics. Because the class size was small and the students actively participated in the discussions, they did not work in groups during class time. However, they were encouraged to work in groups when doing homework.

Pre- and post-tests were administered to determine the amount of growth, if any, experienced by each student and, therefore, each group in the area of a) spatial visualization, b) level-of-understanding in algebra and graphing concepts, and c) spatial visualization and level-of-understanding in each of the three main topics taught during the semester course. These three topics were:

1. Linear Equations
2. Systems of Equations and Inequalities involving Linear Functions with two variables. Absolute Values
3. Parabolas (Vertical only)

Both groups were permitted to use only paper and pencil during the pre- and post-testing sessions.

An example of homework for the traditional group from the text [**17**], pg. 175. is the following:

For each of the following functions, (a) find the vertex; (b) find the line of symmetry; and (c) determine whether there is a maximum or minimum function value and find that value.

1. $f(x) = x^2$ 2. $f(x) = -5x^2$
3. $f(x) = -2(x-9)^2$ 4. $f(x) = 5(x-7)^2$
5. $f(x) = 2(x-1)^2 - 4$ 6. $f(x) = -(x+4)^2 - 3$

In addition, students were asked to determine and state the roles of a, h, and k in the equation $y = a(x - h)^2 + k$.

Worksheets were developed by the instructor to aid students in the use of the graphing calculator while doing the homework. The following is an example from the worksheet for the Calculator Group covering the textbook example given above:

After examining several parabolas of the form

$$y = f(x) = a(x - h)^2 + k$$

answer the following questions.

a)   What happens to the graph of $f(x)$ when:
    1. the value of a increases?
    2. the value of a is positive?
    3. the value of a is negative?
    4. the value of a is positive and between 0 and 1?

b)   What is the line of symmetry and the vertex for each of the graphs?

c)   Under what circumstances does the function result in a maximum or minimum?

Level-of-Understanding Measure in Algebra and Graphs.

The Chelsea Diagnostic Mathematics Tests for Algebra and Graphs (see Figures 3 and 4) were written by the Social Science Research Council Programme 'Concepts in Secondary Mathematics and Science' (CSMS) based at the Centre for Science Education, Kings COllege, University of London. The tests measure and identify a hierarchy of understanding, referred to as level-of-understanding in this paper, connecting key concepts in secondary school [3]. The tests were considered appropriate for this study since many of the topics in algebra and graphs taught in secondary schools are also included in an elementary college algebra course. Because the emphasis during test development was on the understanding of concepts, the tests contain very few items requiring routine mechanical skills or memorized solution techniques [14]. These researchers also established a ladder hierarchy for the test items. To illustrate, the Algebra test consisted of four levels, 1 through 4. Level 1 consisted of the most elementary of the concepts while Level 4 constituted the most advanced of the algebra concepts. The Graphs test consisted of three levels in a similar hierarchy.

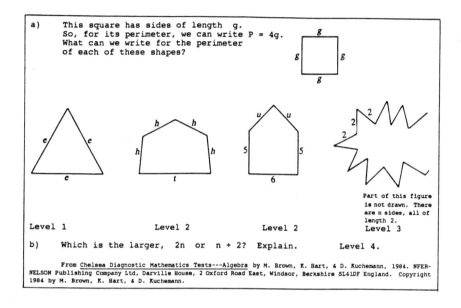

FIGURE 3.   Chelsea Algebra Test

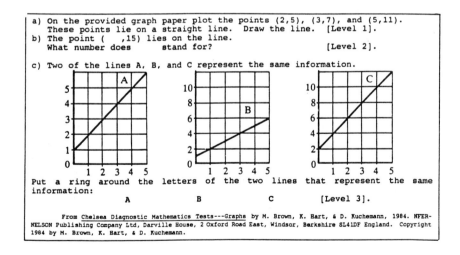

FIGURE 4.   Chelsea Graphs Test

Spatial Visualization and Level-of-Understanding Measure in Each of the Three Selected Course Topics.

While the ETS Tests (see Figures 1 and 2) and the Chelsea Tests (see Figures 3 and 4) served as instruments to measure broad spatial visualization skills and level-of-understanding of elementary algebra concepts, there was also a desire to measure spatial visualization and level-of-understanding specific to those topics presented during the course of study. In each of these three topics, a) linear equations, b) systems of linear equations and inequalities with two unknowns and absolute value, and c) parabolas, a high amount of visualization is required in the presentation, learning and exploration of the concepts. Modeled after the Chelsea Tests, the aim of the six 'mini' tests (three for spatial visualization and three for level-of-understanding) was to examine a student's higher-order understanding in each of the specific topics and the accompanying spatial visualization skills relevant to the topics. The tests were administered in pairs. For example, the Spatial Visualization (SV) test covering Linear Equations was given before that unit of study at the same time as the Level-of-Understanding (LOU) test for Linear Equations. Post-tests were given at the end of each topic of study.

All together nineteen sets of data were examined for each of the 37 subjects to determine the effect of the graphing calculator on spatial visualization skills and level-of-understanding. As listed below, fifteen sets of data were from the individual tests.

> Card Rotation Test
> Paper Folding Test
> Chelsea Algebra Levels #1, #2, #3, and #4
> Chelsea Graphs Levels #1, #2, and #3
> Spatial Visualization (SV) for the 3 course topics
> Level-Of-Understanding (LOU) for the 3 course topics

Four aggregated test combinations were examined as follows:

Chelsea Algebra Levels #1+#2+#3+#4
    = Chelsea Algebra Total
Chelsea Graphs Levels #1+#2+#3
    = Chelsea Graphs Total
Spatial Visualization for the 3 course topics
    = SV Total
Level-of-Understanding for the 3 course topics
    = LOU Total

The data were examined in two fashions. The first obtained the mean and standard deviation for each class for the pre-tests and post-tests. In addition, the gain, i.e., the difference between post-test and pre-test scores, was examined for each student and class. The t-test statistic was calculated to determine the significance of the differences in the gains between the two groups.

The second manner of analysis involved the examination of scattergrams showing the action of the Calculator Group, of the Traditional Group, and of each individual student. This analysis will be illustrated and discussed in the following section of this report.

For many types of research, it would seem quite natural to immediately proceed to a statistical analysis of the data, particularly in cases involving pre- and post-tests. In this study statistics alone would not reveal the entire story of the graphing calculator's effect, particularly for each individual student and her "path" of progress from pre- to post-test time. When dealing with small numbers, it is simple to use scattergrams. One test, Card Rotation, was selected for discussion. Scattergrams for most other tests are also included for reference: those scattergrams for Chelsea Algebra 1, Chelsea Algebra 2, and Chelsea Graphs 1 were excluded because nearly all subjects pre-tested at or near the maximum score.

A method of data presentation was devised that allows the reader to see each student's pre-test score and each student's gain and progress toward a perfect score at the same time.

A discussion of Scattergram 1 for the Card Rotation Test illustrates the information that can be read off all of the scattergrams given in this paper. In each scattergram we let the pre-test score be the abscissa, and gain score the ordinate. Each line of slope -1 passing through the point $(K, 0)$ represents the set of (pre-test, gain) scores corresponding to the fixed post-test score K. Thus it is easy to read off the following two pieces of information from Scattergram 1:

a. each of the five students who gained more than 40 points from pre-test to post-test was a student in the Calculator Group;

b. of the twelve students who scored between 110 and 160 on the post-test, nine were students in the Calculator Group.

The visual details in these scattergrams add much to the understanding of the statistics for the Card Rotation Test.

The scattergram examples deliver a strong case supporting their use as a form of visual data representation. Remaining uncomplicated and uncluttered to the eye, they are packed with information about each group and each individual student. The reader is able to extract enough information from the scattergram to come to conclusions that are also supported by the statistical analysis.

TABLE 1. Descriptive statistics for the fifteen individual tests
(Calculator Group: $n = 19$; Traditional Group: $n = 18$).

Test	pretest	posttest	gain	p-value t-test
Card Rotation (max = 160)				.001
Calculator	86.9(29.25)	113 (31.08)	26.10	
Traditional	91.4(30.57)	97 (30.04)	5.60	
Paper Folding (max = 20)				.035
Calculator	7.40(2.93)	10.40 (3.96)	3.00	
Traditional	9.60(3.38)	10.80(4.47)	1.20	
Chelsea Algebra Level 1 (max = 6)				N.S.
Calculator	5.68(0.48)	5.74(0.65)	.06	
Traditional	5.89(0.32)	5.83(0.38)	-.06	
Chelsea Algebra Level 2 (max = 7)				N.S.
Calculator	5.68(1.08)	6.05(1.31)	.37	
Traditional	6.00(1.14)	6.28(1.23)	.28	
Chelsea Algebra Level 3 (max = 8)				N.S.
Calculator	4.21(1.93)	4.79(2.21)	.58	
Traditional	4.44(2.36)	5.78(2.37)	1.34	
Chelsea Algebra Level 4 (max = 9)				N.S.
Calculator	3.26(2.33)	4.00(2.19)	.74	
Traditional	3.61(2.48)	3.78(2.34)	.17	
Chelsea Graphs Level 1 (max = 7)				N.S.
Calculator	6.47(0.84	6.58(0.61)	.11	
Traditional	6.00(1.28)	6.28(1.18)	.28	
Chelsea Graphs Level 2 (max = 6)				.001
Calculator	3.84(1.61)	5.53(0.84)	1.69	
Traditional	4.39(1.88)	4.50(2.07)	.11	
Chelsea Graphs Level 3 (max = 11)				.015
Calculator	4.74(4.05)	8.68(3.16)	3.94	
Traditional	6.78(3.56)	7.89(3.50)	1.11	

Test	pretest	posttest	gain	p-value t-test
Level-of-Understanding for Linear Equations (max = 13)				.050
Calculator	5.26(2.38)	9.53(3.36)	4.27	
Traditional	7.18(3.45)	9.94(3.69)	2.76	
Level-of-Understanding for Systems (max = 11)				N.S.
Calculator	3.47(2.86)	5.90(2.96)	2.43	
Traditional	4.56(3.31)	5.33(3.60)	.77	
Level-of-Understanding for Parabolas (max = 9)				N.S.
Calculator	1.84(1.84)	6.68(1.83)	4.84	
Traditional	2.17(2.28)	6.44(3.00)	4.27	
Spatial Visualization for Linear Equations (max = 13)				.010
Calculator	6.53(2.32)	9.95(2.59)	3.42	
Traditional	8.22(3.04)	9.61(3.52)	1.39	
Spatial Visualization for Systems (max = 8)				N.S.
Calculator	2.84(2.65)	3.53(2.01)	.69	
Traditional	3.06(2.16)	4.78(3.21)	1.72	
Spatial Visualization for Parabolas (max = 10)				.005
Calculator	3.11(3.04)	8.47(1.51)	5.36	
Traditional	2.50(2.12)	5.17(2.38)	2.67	

TABLE 2. Descriptive statistics for
the four aggregated test combinations

Test	pretest	posttest	gain	p-value t-test
Chelsea Algebra Total (max = 30)				N.S.
Calculator	18.84(4.54)	20.58(5.12)	1.74	
Traditional	19.94(5.56)	21.67(5.17)	1.73	
Chelsea Graphs Total (max = 24)				.002
Calculator	15.05(5.66)	20.79(3.74)	5.74	
Traditional	17.17(5.52)	18.67(5.43)	1.50	
Level-of-Understanding Total (max = 33)				.030
Calculator	10.58(6.28)	22.11(6.61)	11.53	
Traditional	13.89(7.80)	21.72(8.90)	7.83	
Spatial Visualization Total (max = 31)				.023
Calculator	12.47(6.23)	21.95(4.70)	9.48	
Traditional	13.78(6.13)	19.56(8.30)	5.78	

Scattergram 1

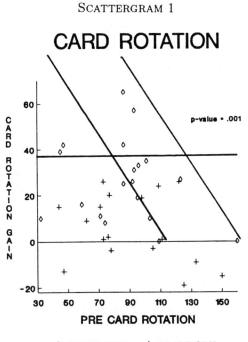

CARD ROTATION

p-value = .001

CARD ROTATION GAIN

PRE CARD ROTATION

◊ CALCULATOR   + TRADITIONAL

CG CARD ROTATION MEAN GAIN = 26.11
TG CARD ROTATION MEAN GAIN = 5.59

Scattergram 2

PAPER FOLDING

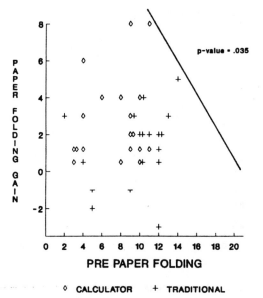

p-value = .035

PAPER FOLDING GAIN

PRE PAPER FOLDING

◊ CALCULATOR   + TRADITIONAL

CG PAPER FOLDING MEAN GAIN = 3.00
TG PAPER FOLDING MEAN GAIN = 1.22

SCATTERGRAM 3

ALGEBRA 3

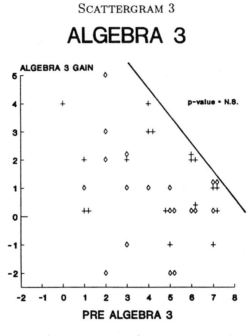

◊ CALCULATOR   + TRADITIONAL

CG ALGEBRA 3 MEAN GAIN • 0.58
TG ALGEBRA 3 MEAN GAIN • 1.34

SCATTERGRAM 4

ALGEBRA 4

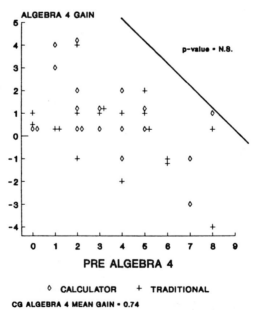

◊ CALCULATOR   + TRADITIONAL

CG ALGEBRA 4 MEAN GAIN • 0.74
TG ALGEBRA 4 MEAN GAIN • 0.17

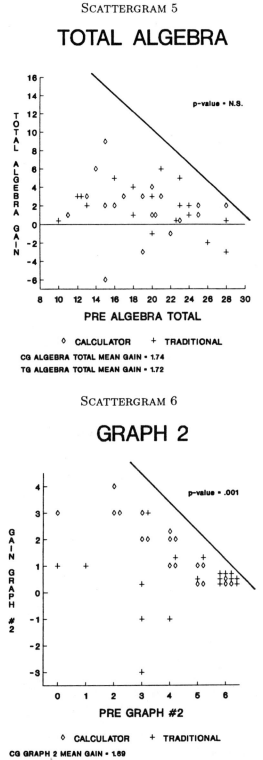

SCATTERGRAM 5

TOTAL ALGEBRA

p-value · N.S.

◇ CALCULATOR   + TRADITIONAL

CG ALGEBRA TOTAL MEAN GAIN · 1.74
TG ALGEBRA TOTAL MEAN GAIN · 1.72

SCATTERGRAM 6

GRAPH 2

p-value · .001

◇ CALCULATOR   + TRADITIONAL

CG GRAPH 2 MEAN GAIN · 1.69
TG GRAPH 2 MEAN GAIN · 0.11

GRAPH 3

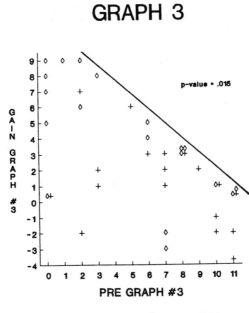

◇ CALCULATOR   + TRADITIONAL

CG GRAPH 3 MEAN GAIN = 3.94

TG GRAPH 3 MEAN GAIN = 1.11

TOTAL GRAPH

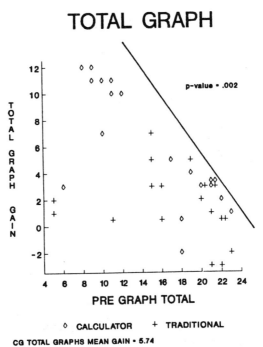

◇ CALCULATOR   + TRADITIONAL

CG TOTAL GRAPHS MEAN GAIN = 5.74

TG TOTAL GRAPHS MEAN GAIN = 1.5

SCATTERGRAM 9

PRELOU 1

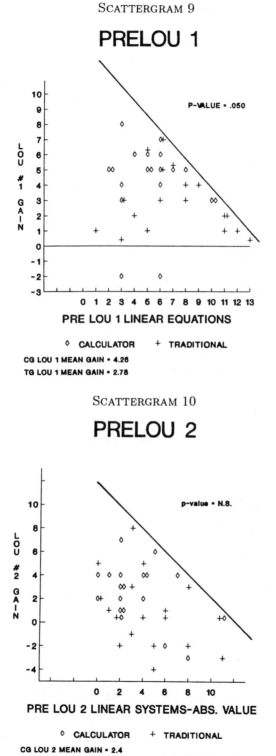

P-VALUE • .050

LOU #1 GAIN

PRE LOU 1 LINEAR EQUATIONS

◊ CALCULATOR   + TRADITIONAL

CG LOU 1 MEAN GAIN • 4.26
TG LOU 1 MEAN GAIN • 2.78

SCATTERGRAM 10

PRELOU 2

p-value • N.S.

LOU #2 GAIN

PRE LOU 2 LINEAR SYSTEMS-ABS. VALUE

◊ CALCULATOR   + TRADITIONAL

CG LOU 2 MEAN GAIN • 2.4
TG LOU 2 MEAN GAIN • .78

SCATTERGRAM 11

PRELOU 3

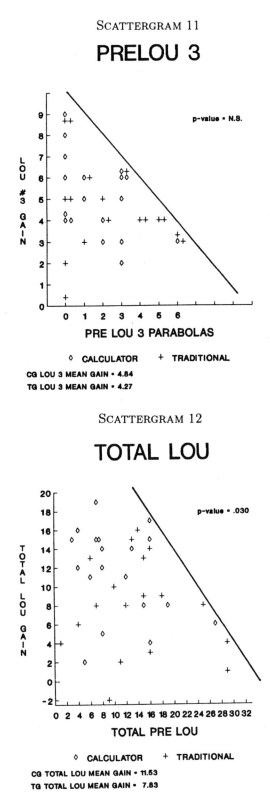

SCATTERGRAM 12

TOTAL LOU

SCATTERGRAM 13

PRESV 1

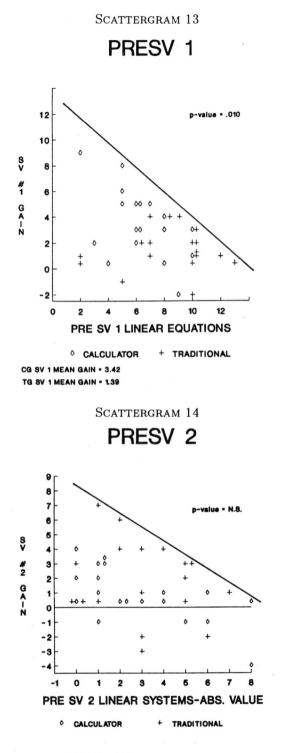

p-value = .010

SV #1 GAIN

PRE SV 1 LINEAR EQUATIONS

◊ CALCULATOR   + TRADITIONAL

CG SV 1 MEAN GAIN = 3.42
TG SV 1 MEAN GAIN = 1.39

SCATTERGRAM 14

PRESV 2

p-value = N.S.

SV #2 GAIN

PRE SV 2 LINEAR SYSTEMS-ABS. VALUE

◊ CALCULATOR   + TRADITIONAL

CG SV 2 MEAN GAIN = 0.69
TG SV 2 MEAN GAIN = 1.72

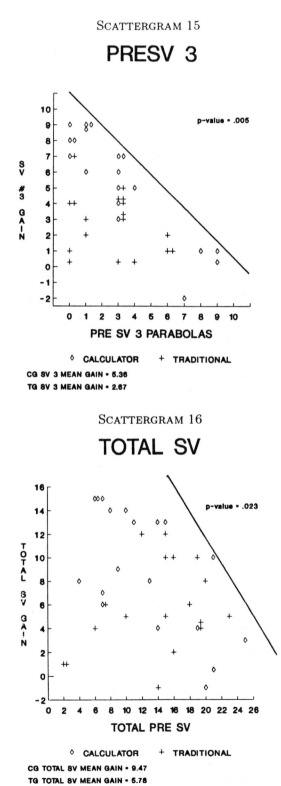

SCATTERGRAM 15

PRESV 3

p-value = .005

CALCULATOR + TRADITIONAL

CG SV 3 MEAN GAIN = 5.36
TG SV 3 MEAN GAIN = 2.67

SCATTERGRAM 16

TOTAL SV

p-value = .023

CALCULATOR + TRADITIONAL

CG TOTAL SV MEAN GAIN = 9.47
TG TOTAL SV MEAN GAIN = 5.78

DISCUSSION OF RESULTS

The statistics in Tables 1 and 2 support the pictorially represented data in Scattergrams 1 and 2. The significance of the gains in spatial visualization ability made by the Calculator Group supports prior research concluding that these skills can be taught. Students use these strengthened abilities to mentally move or alter all or part of a function representation. The means of enhancement in this case was the graphing calculator – a tool more readily available to students than computer graphics. The use of the graphing calculator is easily incorporated into the curriculum and follows the recommendations of researchers urging educators to encourage the development of visual-spatial experiences for women and girls.

Although the individual and aggregated results for the Chelsea Algebra given in Tables 1 and 2 and Scattergrams 3, 4, and 5 do not yield significance, they are of interest. Those gains made by the Calculator Group at Levels 1, 2, and 4 were substantial enough to offset the loss in gain at Level 3 when the four levels are combined for consideration as a whole. After pre-testing at a lower mean at all four levels, the Calculator Group pulled even with the Traditional Group in the post-test which can be seen in Scattergram 5.

The implications of the Chelsea Graphs Total statistics are particularly note-worthy when recalling that the test items were developed to minimize the need for routine mechanical skills and memorized techniques. Therefore, the concepts under investigation were often presented in unfamiliar form. It was at the higher level of test items that the Chelsea research indicated a large gap between the relatively simple reading of the information from a graph and the algebraic expression [19]. These conclusions are supported by the statistics in Tables 1 and 2 and are visually seen in Scattergrams 6 and 7. The Traditional Group did not show strength in understanding the relationship between a graph and its algebraic expression which is required at Levels 2 and 3 [19]. This caused smaller gains for the Traditional Group as the test items became more difficult even though this group started with higher pre-test means at Levels 2 and 3. The strength of the gains on the part of the Calculator Group at Levels 2 and 3 is reinforced when the three levels are considered as one testing instrument. These statistics are given in Table 2 with Scattergram 8 showing the visual interpretation. The Calculator Group's gain of 5.74 points was more than three times the 1.5 point gain by the Traditional Group. Five of those using the graphing calculator achieved perfect post-test scores against two from the non-calculator group.

The Level-of-Understanding (LOU) test results given in Tables 1 and 2 and Scattergrams 9, 10, 11 (the three topics), and 12 (Total) show the Calculator Group pre-testing at a lower mean in each test. Although this group post-tested with higher means in two of the three topics and made greater gains in all test cases, the onlly one of significance was for Linear Equations LOU in favor of the Calculator Group. Systems of equations and inequalities and absolute values were difficult concepts as evidenced by low post-test means for both groups.

When the three topics were put together and treated as one testing instrument, the LOU Total post-test means were still considered low, with each group scoring approximately 22 points out of the total 33. However, the t-test showed significance in favor of the Calculator Group whose mean gain was nearly 12 points compared with approximately 8 points for the Traditional Group. A possible conjecture would be that those using the graphing calculator were able to understand more about the concepts of each topic and then build upon that knowledge to use at the next level.

The Spatial Visualization (SV) data for each topic show the Calculator Group pre-testing at a lower level in two of the three topics as can be seen in Scattergrams 13, 14, 15 (the three topics), and 16 (Total) and post-testing at a lower level in one topic. Significance was in favor of the Calculator Group for SV Linear Equations, SV Parabolas, and SV Total.

The wide gap between the two groups in spatial ability on parabolas is particularly interesting when taken in context with the scores and gains of Spatial Visualization 2 which included absolute values. [As noted in the research log: Both groups had found absolute values to be difficult, but the Calculator Group verbalized the fact that absolute values can be "smoothed out" to make parabolas. Those concepts they had struggled to understand about absolute values could now be "transferred" to parabolas!] In other words, the Calculator Group grasped the similarities between absolute values and parabolas by using their past experiences with the absolute values functions. The Traditional Group failed to make this transfer from absolute values to parabolas and ended with weaker visual thinking scores. The results of the third test covering parabolas suggest that, while the Calculator Group took longer to assimilate the absolute value concepts, it appears they built a stronger conceptual basis which enabled them to elevate to a higher level of abstraction involving parabolas. In this case the Traditional Group did not build as firm foundation at the earlier level of knowledge. The transfer and abstraction of this knowledge then occurred naturally for the Calculator Group, as is shown in their gain mean for parabolas given in Scattergram 15.

CONCLUSIONS

It is conjectured that the reasons for the wide gap between the two groups center upon the capabilities of the graphing calculator. Mathematics educators want their students to understand functions and their graphical, algebraic, and numerical representations. Harel and Dubinsky [13] argue that function "is the single most important concept from kindergarten to graduate school and is critical throughout the full range of education" (p. vii). The mastery of elementary functions includes understanding these three types of representations, any one of which should 'connect' into either of the other two. However, the research shows that those in the Traditional Group were not as successful in making these connections. If one thinks of a 'loop of knowledge' that can be entered from any point, then given the algebraic representation of a function, the student should be able to generate the numerical table and graph of that function. Or, given the graph, the student should be able to 'see'enough information in that graph to enable her to arrive at the algebraic and numerical representations of the function [36, 40].

Moschkovich, Schoenfeld, and Arcavi [24] maintain that students must be able to make connections across the numerical, algebraic, and graphical representations of a function. They describe two different perspectives regarding functions. The 'process perspective'is one in which functions are interpreted as a linking of x- and y-values. The 'object perspective'is one in which function and any of its representations become entities. Schwartz [28] has stated:

> ... symbolic representation of function makes its process nature salient, while the graphical representation suppresses the process nature of the function and makes the function into an "entity."A proper understanding of algebra requires that students be comfortable with **both** of these aspects of function (p. 4).

The graphing calculator gives the student the power to tackle this process of making connections at her own pace. It provides a means of concrete imagery that gives the student new control over her learning environment and over the pace of that learning process. It relieves the need to emphasize symbolic manipulation and computational skills and supports an active exploration process of learning and understanding the concepts behind the mathematics. Numbers are still important, but not to the extent as before. Conceptual understanding takes the place of rote memorization. The student builds concept knowledge from past experiences. The pace of this building process is controlled by the student enabling her to build a solid understanding of the concept which could then lead into abstraction of the concept.

Personal observation showed that the graphing calculator provides a concrete mechanism that can aid the less able student to discover hypotheses on her own, state and re-define those hypotheses, and then projeect the confidence to argue the strength of her hypotheses with her peers. This leads to higher-order-thinking-skillls which translate to higher levels-of-understanding for the student.

This form of learning can be described as an interactive conversation between the student and the graphing calculator. Indeed, the calculator reesponds only to the prompts of the student — it does not act on its own. The student at first silently converses with herself through the calculator, asking questions of herself as she manipulates the concrete screen image. This leads her to have the knowledge and conficence to conjecture not only what is actually occurring with the image, but why it is happening. By this point she is ready to state her conjectures and reasoning supporting her conclusions. This leads into verbal communication with her peers and instructor. The graphing calculator supports this process of 'communicating with oneself'in a manner not available before.

The graphing calculator's visual representation of the mathematical concepts is important as both a heuristic and a pedagogic tool – particularly for those weaker students in mathematics. It is conjectured that through this use of a visual, self-paced exploration process the student is able to use concrete imagery to move toward a higher level of understanding or abstraction. Von Glasersfeld [39] stated that mathematical knowledge is a product of conscious reflection — the more abstract the concept, the more reflection needed. By using the graphing calculator students are more likely to construct their own mathematical understanding through conscious reflection. Spatial visualization, or, if preferred, visual thinking (the term used by the NCTM) should be nurtured and developed in students. It plays a significantly important role in the development of mathematical reasoning. Because this study indicates that the graphing calculatore improves females'spatial visualization, and spatial visualization is an important spatial skill, it can be argued that the graphing calculator could also improve spatial visualization in males. Furthermore, visual thinking promotes the creative and insightful use of mathematical concepts that can transfer to other areas of learning. Students become actively involved in the discovery and understanding process, no longer viewing mathematics as simply the receiving and remembering of algorithms and methods of solution.

Research, curriculum design, and learning environments should be focused on ways to help students increase their spatial visualization skills and redesigned to promote gender equality and encourage the development of visual-spatial experiences for women and girls [21]. To emphasize the connections between the various representations of mathematical concepts, educators should encourage students to use the power of visualization in discovering and understanding mathematics. Each concept representation taken by itself does not give the whole picture of the problem situation. The emphasis on the connections between the various representations will help students to understand through visualization [39] and build a stronger platform upon which to further study mathematics. Readily available and realistic in terms of budget constraints, the graphing calculator is a means by which these goals can be realized in schools and colleges. The results of this study conclude that the graphing calculator does enhance females'spatial visualization skills and level-of-understanding in elementary graphing and algebra concepts.

REFERENCES

1. Baenninger, M. & Newcombe, N., *The role of experience in spatial test performance: A meta-analysis.*, Sex-Roles **20** (1989), 327-344.
2. Ben-Chaim, D., Lappen, Gl, & Houang, R., *The effect of instruction on spatial visualization skills of middle school boys and girls*, American Education Research Journal **25** (1988), 51-71.
3. Brown, M., Hart, K., & Kuchemann, D., *Chelsea diagnostic mathematics tests*, NFER-Nelson, Publishing Company, Windsor, Berkshire, England, 1985.
4. Cartledge, C. M., *Improving Female Mathematics Achievement*, ERIC Document Reproduction Services No. ED 250 198, 1984.

5. Collis, B., *Research retrospective: 198501989*, The Computing Teacher **16(9)** (1989), 5–7.

6. Connor, J. M. & Serbin, L. A., *Mathematics, visual-spatial ability, and sex roles. (Final Report)* (1980), National Institute of Education (DHEW), Washington, D. C. (ERIC Document Reproduction Services No. ED 205 305.

7. Demana, F., & Waits, B. K., *The Ohio State University Calculator and Computer Precalculus Project: The mathematics of tomorrow today!*, The AMATYC REVIEW **10(1)** (1990), 46-55.

8. Dunham, P. H., *Mathematical confidence and performance in technology enhanced precalculus: Gender-related differences. Doctoral dissertation, The Ohio State University, 1990* (1991), Dissertation Abstracts International, 51, 3353A.

9. Ekstrom, R. B., French, J.W., & Harmon, H. H., *Manual for kit of factor-referenced cognitive tests.*, Educational Testing Service, Princeton, NJ, 1976.

10. Fennema, E., *Spatial ability, mathematics and the sexes*, Mathematics learning: What research says about sex differences: E. Fennema (Ed.),, ERIC Clearinghouse for Science, Mathematics, and Environmental Education.

11. Fennema, E. & Tartre, L., *The use of spatial visualization in girls and boys*, Journal for Research in Mathematics Education **16** (1985), 184–206.

12. Flores, Al, & McLeod, D., *Calculus for middle school teachers using computers and graphing calculators Paper presented at the Third Annual Conference on Technology in Collegiate Mathematics, Columbus, OH.* (1988).

13. Harel, G., & Dubinsky, E., The concept of function–Aspects of epistemology and pedagogy G. Harel & E. Dubinsky, (Ed.), vol. 25, Mathematical Association of America Notes, Washington, D.C., 1992.

14. Hart, K., Brown, M., Kerslake, D., Kuchemann, D., & Ruddock, J., *Children's understanding of mathematical concepts*, John Murray, London, 1981.

15. Horowitz, L., *Visualization and aithmetic problem-solving. Paper presented at the American Educational Research Association Annual Meeting* (1981).

16. Hyde, J.S., Fennema, E. & Lamon, S., *Gender differences in mathematics performance; A meta-analysis*, Psychological Bulletin **107** (1990), 139-155.

17. Keedy, M.L., & Bittinger, M.L., *College algebra: A functions approach*, Addison Wesley Publishing Company, Reading, MA, 1986.

18. Kersh, M.E., & Cook, K. H., *Improving mathematics ability and attitude, a manual*, Mathematics Learning Institute, University of Washington, Seattle, WA, 1979.

19. Kerslake, D., *Graphs*, Children's understanding of mathematics, John Murray, London, 1981, pp. 11–16.

20. Lappan, G., & Phillips, E., *The mathematical preparation of entering college freshmen*, NASSP-Bulletin **68** (1984), 79-84.

21. Linn, M. & Hyde, H.J., *Gender, mathematics, and science.*, Educational Researcher **18(8)** (1989), 17–19.

22. Mathematical Association of America, *Challenges for college mathematics: An agenda for the next decade. Report of a Joint Task Force of the Mathematical Association of America and the Association of American Colleges.* (1991), Association of American Colleges, Washington, D.C..

23. McGee, M. G., *Human spatial abilities: Sources of sex differences* (1979), Praeger Publishers, New York.

24. Moschkovich, J., Schoenfeld, A. H., & Arcavi, A., Integrating research on the graphical representation of function, T. A. Romberg, E. Fennema, and T. P. Carpenture (Eds.), Erlbaum, Hillsdale, NJ, 1992.

25. National Council of Teachers of Mathematics, *Curriculum and evaluation standards for school mathematics* (1989), Author, Reston, VA.

26. Newcombe, N., *Sex differences in spatial ability and spatial activities*, Sex-Roles: A Journal of Research **9** (1983), 377-386.

27. Ruthven, K., *The influence of graphic calculator use on translation from graphic to symbolic forms*, Educational Studies in Mathematics **21** (1990), 431–450.

28. Schwartz, B., *Functions and graphs*, The concept of function: Aspects of epistemology and pedagogy G. Harel & E. Dubinsky, (Ed.), vol. 25, Mathematical Association of America Notes, Washington, D.C., 1992.

29. Sherman J., *Factors predicting girls' and boys' enrollment in college preparatory mathematics*, Psychology of Women Quarterly **7(3)** (1983), 272-281.

30. Silver, E. A., Kilpatrick, J., & Schlesinger, B., *Thinking through mathematics*, College Entrance Examination Board, New York, 1990.

31. Tartre L.A., *Spatial orientation skill and mathematical problem solving*, Journal for Research in Mathematics Education **21** (1990), 216-229.

32. Tartre, L. A., *Spatial skills, gender and mathematics*, Mathematics and gender, G. Leder and E. Fennema (Eds.), Teachers College Press, Teachers College, Columbia University, New York and London, 1990.

33. Taylor L. J. C., *Assessing the graphing levels of understanding and quadratic knowledge of $C^2 PC$ students*, Proceedings of the Second Annual Conference on Technology in Collegiate Mathematics F. Demana, B. Waits, and J. Harvey (Eds.), 1990, pp. 324–327.

34. Threadgill-Sowder, J., A., & Julifs, P. A. Manipulative versus symbolic approaches to teaching logical connectives in junior high school: An aptitude X treatment interaction study, Journal for Research in Mathematics Education **5** (1980), 367–367.

35. Tohidi, N. E., *Gender differences in performance on tests of cognitive functioning: A meta-analysis of research finding. paper based on doctoral dissertation, University of Illinois* (1986).

36. Trotter, A., *Graphing Calculators are coming to class*, Executive Educator **13** (1991), 20–31.

37. Vasquez, J. L., *The effect of the calculator on student achievement in graphing linear functions (Doctoral Dissertation, University of Florida, 1990*, Dissertation Abstracts International **50, 3508A** (1991).

38. Vonder Embse, C., *Visualization in precalculus: Making connections using parametric equations*, Monograph by the Department of Mathematics, Central Michigan University (1991).

39. Von Glasersfeld, E., *Learning as a constructive activity*, Problems of representation in the teaching and learning of mathematics, C. Janvier, (Ed.), Erlbaum, Hillsdale, NJ, 1987, pp. 3–17.

40. Waits, B. K., & Demana, F., *A case against computer symbolic manipulation in school mathematics today*, Mathematics Teacher **85** (1992), 180–183.

COLLEGE OF NEW ROCHELLE

CBMS Issues in Mathematics Education
Volume **4**, 1994

To the Right of the "Decimal" Point: Preservice Teachers' Concepts of Place Value and Multidigit Structures

RINA ZAZKIS & HELEN KHOURY

This study is concerned with how college students, in specific, preservice elementary school teachers, understand the concepts related to the positional number systems, such as multidigit structures, rational number representation and place value. The positional sexagesimal (base 60) number system that included fractional numbers was invented and used by the Babylonians as early as 2000 BC. Probably, this invention was premature for the needs of humanity and therefore it wasn't disseminated and accepted outside of Mesopotamia. The genuine *decimal* place value number system for integers was conceptualized in India in the sixth century A.D. Decimal fractions were independently invented several times, but they came to general use only at the end of sixteenth century [**11**]. It took almost ten centuries for the humankind to expand the idea of place value to the fractional part of a number. We would like to emphasize not the date of invention of the positional numeration system concept, but the period of its cultural acceptance.

A late historical acceptance of a concept, as for example in the case of negative or irrational numbers, usually indicates some conceptual or psychological difficulty for a learner. Are we aware of students' difficulties with the place value representations in the fractional part of the number? Do we acknowledge these difficulties in the curriculum and in teacher preparation courses? Due to daily use in a variety of settings, as well as schooling experiences, individuals have learned to manipulate decimal fractions with reasonable success and comfort. In this study we were interested to examine teachers' understanding beyond algorithmic manipulations.

The general concept this study investigates is the concept of place value, emphasizing the place value in a fractional part of a rational number. More specific concept of our concern is a non-integer number represented in bases other than ten. These number representations, such as 12.34_{five} or 46.23_{seven}, are referred to as "non-decimals". A detailed analysis of the domain of non-decimals, including comparison of non-decimals to decimal fractions, can be found in Zazkis and Khoury [26]. Various researchers (e.g. [17, 20, 23]) have pointed out that a correct algorithmic performance in a given domain doesn't necessarily indicate students' conceptual understanding. Therefore, assuming teacher's ability to manipulate correctly decimal rational numbers, we asked them to make sense of "non-decimals" by performing addition and subtraction with these numbers and also by converting them to base ten representation. These non-standard tasks helped to eliminate the possibility of students' rote learned patterns and assisted in focusing on place value. Students' efforts to interpret "non-decimals" help us find possible deficiencies in their understanding of concepts of rational numbers and multidigit structures in base ten.

The conventional way of converting 12.35_{five} to base ten is to use the following expanded notation:

$$12.34_{\text{five}} = 1 \times 5 + 2 \times 1 + 3 \times \frac{1}{5} + 4 \times \frac{1}{25} = 7\frac{19}{25} = 7.76_{\text{ten}}.$$

Zazkis & Khoury [26] categorized students' problem solving strategies and possible errors or misconceptions in interpreting "non-decimals," based on a group administered written assessment. The current study is based on the analysis of clinical interviews and it attempts to provide a conceptual framework for student's mental constructions in this domain. Our objectives were to investigate deficiencies in students' understanding of place value, using non-decimal representation as a research tool, and to map learner's concepts of place value and multidigit number structure.

THEORY

Action-process-object framework.

Our theoretical perspective is a constructivist approach, based on the ideas of Piaget. The particular interpretation of constructivism used in this study is the action-process-object developmental framework suggested by Dubinsky (e.g. [7, 8]). Dubinsky discusses the action-process-object theoretical perspective as an adaptation of ideas of Piaget to the studies of *advanced* mathematical thinking. Previously this theory was used in the studies of undergraduate mathematics topics like calculus and abstract algebra (e.g. [2, 3, 7, 9, 10]). We suggest that this theoretical perspective is appropriate for the discussion of mathematical knowledge development in general, not necessarily of what is considered to be "advanced." One of the goals of this study was to examine this claim.

The essence of the theoretical perspective developed by Dubinsky is that an individual, disequilibrated by a perceived problem situation in a particular social context, will attempt to reequilibrate by assimilating the situation to existing schemas available to her or him, or, if necessary, to reconstruct those schemas at a higher level of sophistication. The constructions which may intervene are mainly of three kinds — actions, processes, and objects.

An *action* is any repeatable physical or mental manipulation that transforms objects in some way. When the total action can take place entirely in the mind of an individual, or just be imagined as taking place, without necessarily running through all of the specific steps, we say that the action has been *interiorized* to become a *process*. New processes can also be constructed by inverting or coordinating existing processes. When it becomes possible for a process to be transformed by some action, then we say that it has been *encapsulated* to become an *object*.

As an example of a not-so-advanced mathematical concept, we consider the number 5 and describe the development of this concept using an action-process-object framework. In the beginning of number acquisition the number 5 is identified with the counting of five attributes [18]. A young learner has to touch the things counted to build a one-to-one correspondence between the number of attributes and the counting sequence. At this level the number 5 is an action for the learner. Later, when the learner can count a group of five in her or his mind, or recognize a group of five , we would say that the number 5 has been interiorized to become a process. When a learner is able apply an action to the number 5, such as adding it to another number, we say that number 5 has been encapsulated to become an object.

In many mathematical situations it is essential to be able to shift from an object back to a process. One of the tenets of the theory is that this can only be done by *de-encapsulating* the object, that is, to go back to the process which was encapsulated in order to construct the object in the first place. In our example de-encapsulating of the object of number 5 may take place when the learner is asked to compare number five with number six. In this case the learner may go back to the process of counting to discover that number 5 is reached in the counting sequence before the number 6, therefore it is smaller. The concept of de-encapsulation will play a major role in the analysis of our data.

In what follows we describe a *genetic decomposition* [8, 9] for the concept of fractional number represented in bases other than ten, that is, the specific constructions that can be made in order to understand a particular topic of non-decimal representations of non-integer numbers. We also describe a genetic decomposition for addition/subtraction of these numbers.

Genetic decomposition for a conversion task.

By "conversion" here we mean conversion from "non-decimals" in various bases to base ten representation. The task of conversion from common or decimal fractions to non-decimal representations, in the general case, requires application of concepts from number theory, such as prime decomposition, relatively prime numbers and the use of Fermat's Little Theorem; concepts which are not in the scope of knowledge of the participants of this study. This is a challenging mathematical problem, that a reader is invited to explore or to consult Zazkis and Whitkanack [27].

Let us consider the number 12.34 in base-five. The problem situation of converting this number to base-ten representation is a source of disequilibration, since it presents a novelty within familiar symbols. The novelty is caused by unfamiliar coordination of symbols and concepts. The symbol 12.34 is familiar as well as the concept of representation of integers in different bases. The concepts of decimal fraction representations and other-than-ten bases representations need to be coordinated in the student's mind to create a new concept of "non-decimal."

There are (at least) two possible ways to construct an understanding of non-integer numbers represented in other-than-ten bases. Using 12.34_{five} as an example, one possibility to construct an understanding of this number representation is to imitate, to draw an analogy with the mental construction of 12.34 in base ten. One common approach is to start with constructing an understanding of base-five integers and common fractions, and then to proceed to "pentomal" fractional representations as a way of recording fractions with denominator of five and powers of five.

Another route that a learner may go through in the presented problem situation is to start with the familiar mental object of 12.34_{ten} and to reconstruct this object to include the possibility of being 12.34_{five} or, in general, 12.34_{anybase}. How may 12.34 be reconstructed to become a specific example of 12.34_{anybase}? The key word here is *de-encapsulation*, the de-encapsulation of an object 12.34_{ten} to a process of 12.34_{ten}. The process of 12.34_{ten} is best represented in its expanded notation, that is $1 \times 10^1 + 2 \times 10^0 + 3 \times 10^{-1} + 4 \times 10^{-2}$. Similarly, the process of 12.34_{five} is represented by $1 \times 5^1 + 2 \times 5^0 + 3 \times 5^{-1} + 4 \times 5^{-2}$ and the process of $12.34_{\text{(base B)}}$ is represented by $1 \times B^1 + 2 \times B^0 + 3 \times B^{-1} + 4 \times B^{-2}$, where B stands for any integer base. Then this process may be encapsulated to a more sophisticated object of place value representation of 12.34.

The first possibility discussed above is analogous to learning Greek by being in Greece, that is, learning a foreign language in its natural environment. This is, of course, if we agree on the analogy that base-ten representation is the "mother tongue," the first language. The second possibility discussed above is analogous to learning Greek by constantly translating the meaning to and from English (assuming English is the first language). It is not surprising that the students in this study preferred the second approach and variations on it; after all, telling a story about the Yellow Planet ([21], pp. 29–34) where base five is a convention, is different from experiencing living on it.

Genetic decomposition for an addition/subtraction task.

Addition/subtraction of two non-decimals can be done in two possible ways: working directly in the new base and working with constant reference to base ten. For example, summing up 4 and 3 in base 5 directly, one may recall the "basic fact" that $4_{\text{five}} + 3_{\text{five}} = 12_{\text{five}}$. Those who don't have basic addition facts for base five in their active memory, may derive this fact, for example, by counting up: 4, (5-oops–10)10, 11, 12. When making reference to base ten, one would perform the addition of 4 and 3 in base 5 by recalling that $4 + 3 = 7$ in base ten, and then representing the number 7 as 12_{five}.

A metaphor of building a sentence in a foreign language — once by trying to think in this foreign language and the second time by translation from the familiar language — is, again, a powerful analogy to explain the difference between the two approaches.

METHOD

Subjects.

Twenty university students participated in the study. They were pre-service elementary school teachers in the last year of their teacher training, participating in the "methods" course for teaching elementary school (K-8) mathematics. All of them had a course in "Foundations of mathematics for teachers" in their background that included the topic of numbers in different numeration systems, specifically, number representations in bases other than ten. The topic of number representations in different bases was revisited during the methods class. This instructional treatment included the use of base block manipulatives and was designed to help preservice teachers understand the difficulties that their elementary students may have with the ideas of place value representation in base ten. However, the idea of extending other-than-ten base representation to the fractional part of the number was not introduced prior to the administration of the interviews.

Instrument.

In a clinical interview setting students were presented with two different types of problems. In the first type of problems the students were asked to perform addition and subtraction with non-decimals. In the second, the students were asked to convert non-decimals to base ten. As the first step in the interview, that served as a screening device, students were given similar tasks involving integers in various bases, but not non-decimals. During each interview, that lasted about one hour, participants were prompted where appropriate for understanding, that might not have been apparent from their initial response. Clarification and probing questions were asked, additional tasks were offered where necessary, to establish or to confirm the strategy used by a student. The interviews were transcribed and categorized in terms of different questions, their difficulty, and repeating error patterns used by students. The action-process-object theoretical perspective was used to analyze the interviews for ways in which the students appeared to think about the specific topics.

RESULTS

All the participants passed a screening assessment, that is, they were able to perform correctly on tasks of conversion from various bases to base ten and of addition/subtraction with integers, represented in bases other than ten. With non-integer numbers the situation was different. Sixteen out of twenty participants were able to perform addition/subtraction correctly using different reasoning strategies. Four other students implemented "error patterns," some of which were consistent with some of the examples described by Ashlock and Van-Lehn [1, 23]. Only 10 students out of 20 converted the given non-decimals to base ten representation correctly. The remaining ten were consistent in using self-invented incorrect methods through different examples. In what follows we describe and exemplify students' responses to the proposed tasks.

Addition/Subtraction Tasks.

Among the students who succeeded in the addition/subtraction tasks, the majority (13 out of 16) preferred to make consistent reference to base ten calculations. The other three students attempted to derive basic addition/subtraction facts for the base in question without using their knowledge of the basic facts in base ten.

Addition/subtraction with reference to base ten.

Interviewer:

Could you add 2.34 and 4.32 in base 5?

Ann:

Okay. Well, I guess because I'm so familiar with base 10, no matter what I do I'll probably think of it in base 10 and convert to base 5. It's like when you're doing any language for the first time, you're still converting to what your, your own, your native language, or whatever you want to call it is. So I would go 4, I'm adding, right? Okay, so 4 plus 2 is 6, but there's no 6 in base 5, so I say 1 carry the 1. And then 3 and 3 is 6 plus 1 is 7, but there's no such thing as 7 so 2 carry the 1. And then there's 7 but there's no such thing so it's a 2, like that.

Interviewer:

Okay, how about something like 6.54 subtract 4.56 in base 7?

Ann:

Okay , you can't take 6 away from 4 so I would borrow and cross it out and then I would give myself, so that's now 4 and 1 of these, that's 11. Now 11 minus 6 is 5 and then I need to borrow again so that's 14, no, I can't actually, 7 and 4 is 11, 11 minus 5 is 6, so 1.65.

Interviewer:

How did you come up with this procedure?

Ann:

It, it's similar to the procedure that I use when I'm using base 10. , I'm still borrowing 1 unit from the number to the left when I'm subtracting and I don't have enough of my group to subtract from.

Ann constructs the process of addition/subtraction algorithm for base five using explicitly the familiar process of addition/subtraction algorithm for base ten — "no matter what I do I'll probably think of it in base 10." The new process is constructed by coordinating the familiar process with new "basic facts" for addition, that is, with addition table for base five. Those "basic facts" are also processes for Ann, she derives each one of them by first recalling the sum in base

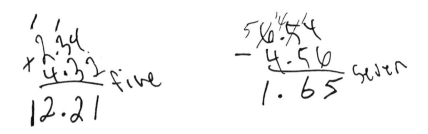

FIGURE 1. Addition and subtraction performed by Ann.

ten and then representing it in base five — "4 plus 2 is 6, but there's no 6 in base 5, so I say 1 carry the 1."

The procedure used by Ann, and replies like "there's no such thing as 7 (in base 5)," are typical for this group of interviewees. In Ann's reasoning, as well in the reasoning of other students, who implemented similar algorithms, we don't find any reference to the "decimal" point or to the explicit place values of the digits to be added. The idea of "borrowing 1 unit from the number to the left," and the ability to interpret the meaning of the "unit" borrowed as "4 and 1 of these, that's 11," demonstrate the awareness of the process of relationship between place values of neighbor digits. However, detecting the place values of specific digits doesn't seem to be necessary for performing addition/subtraction.

Addition/subtraction without reference to base ten.

Ruth mentions the result of calculation in base ten, but doesn't use this reference in her calculations. Instead, she calculates the required sums and differences by counting up and counting down.

Interviewer:

How about 6.42 plus 2.53 in base 7?

Ruth:

Okay. So 3 and 2 would still be 5, 5 and 4 is 9 but you can't do 9, so it would be 5,6,10,11,12? [read: five, six, one-zero, one-one, one-two] I think so. Six and 1 is 10 [read: one-zero] and add 2 is 11,12, [read one-one, one-two] so 12.25?

Interviewer:

Okay, how about if we were to do 6.42 subtract 2.53?

Ruth:

Subtract? Oh no!

Interviewer:

I'm still trying to get some sweat.

Ruth:

(Laugh) You might on this one. Two minus 3 you can't do, so borrow 1 from here, what base is this? Seven?

Interviewer:

Uh, 7.

Ruth:

So borrow 1 from there, but and you just borrow, but borrowing a 10 [read: ten] wouldn't make it 12 [read: twelve], right? Because this isn't a 10, it's the 7th (pause) hmmmm! You'd add, so this should be 9 now because you'd add 7 instead of a 10, so then this 9, but you can't have 9 in base 7, so it would be 11, 12 [read: one-one, one-two]. Oh, that's what I had before. 12 [read: one-two] minus three, so it would be 11,10,6 [read one-one, one-zero, six] yeah, 6 and 3 and 5 you can't do, so you add uh, how do I do it. I just add 10 to the top this. No, you could add 7, because it's base 7, right? You're not going to tell me anyway. Ten (laugh), this is (counting) 11, 12, 13 in base seven. 13 minus 5 so 13, 12, 11, 10, 6, 5, using my fingers, five here. And this is 5 minus 2 is 3, even close (laugh).

Interviewer:

Alright, so you have 6.42 minus 2.53 is uh $\cdots$

Ruth:

3.56$\cdots$.

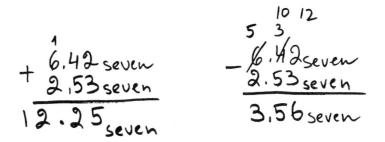

FIGURE 2.   Addition and subtraction performed by Ruth.

It is interesting to note that the simple addition facts like $10 + 1 = 11$ or $10 + 2 = 12$, that are true in <u>any base</u> (> 2) were not taken by Ruth for granted. She had to count up to derive this fact in base seven. When she reached the conclusion by counting up, Ruth mentioned with some surprise "Oh, that's what I had before." This discovery is not being generalized and used in another subtraction task that Ruth performed during the interview — she always counts up or counts down, at times using her fingers.

Both excerpts demonstrate the importance of having "basic facts" as objects and the ability to recall them, of knowledge of the "addition table," for multidigit addition and subtraction. In both cases the students were not able to recall the sum for $(x + y)_{\text{base B}}$, where x and y are single digit numbers. These students haven't yet constructed the object for $(x + y)_{\text{base B}}$ in their mind. For Ruth $(x + y)_{\text{base n}}$ is an action of counting. For Ann $(x + y)_{\text{base B}}$ is a process that is constructed by de-encapsulation of the object $(x + y)_{\text{ten}}$ and reconstructing, regrouping it as $B + [(x + y)_{\text{ten}} - B]$.

As mentioned above, 16 out of twenty students performed the addition and subtraction tasks mostly correctly. Occasional errors in computation are beyond the scope of this article. In what follows, we discuss the systematic strategies used by the other four students, who generated inappropriate algorithms for addition. Each one of the invented computational error patterns is unique in this sample, but was used consistently by the inventor.

Mirror pattern in addition/subtraction.

In performing addition and subtraction Mary separated the integer and the fractional parts of the number. The addition/subtraction of the integer part was performed correctly, starting on the right and adapting the standard algorithm to a different base. The addition/subtraction in the fractional part of the number was treated separately and started on the left. In this strategy the "decimal" point serves as a "mirror," everything that was done in the integer part of the number right to left is reflected in the fractional part of the number left to right. After detecting the strategy, in the excerpt below the interviewer has chosen to involve more digits in the addition problem to make the algorithm used by Mary more explicit.

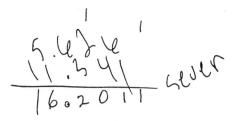

FIGURE 3.   Addition performed by Mary.

Interviewer:

Yeah, I'll give you the next question now. Something like 5.626 plus 11.341 in base seven. Please talk aloud about what you're doing.

Mary:

Okay, now we're only, we're representing in base 7, I'm not representing in, I've got to think here, just a sec., okay, talking out loud here. Okay I'm going to start with the units what we've started with on the left side, we worked on the left side of the decimal. On the left side 5 plus 1 is 6, I don't have any carry overs because it's base 7, that's fine, the 1 can come down, that's fine. And the same on the opposite side of the decimal. And the right hand side of the decimal, 3 plus 6, 9 is not part of that base so I would have a remainder of 2, carry one over to the next column, I would have 3 plus 4 which is the ??, which is 7, I can't keep it there, so that would represent with a 0, and I take 1 whole group over and that would give me 8 which I can't keep, a 1 remainder, take 1 over again, I'd have another, so I'd have 16.2011. That's the way I would look at it anyway.

Mary 's confusion and her meaning for the place values in the "opposite side" will be addressed further when we discuss her conversion strategy. Theoretical discussion of Mary's computational algorithm is provided following the next excerpt.

Confusion with carry pattern in addition.

Confusion between the digit to write and digit to carry has been described in the earlier research with children learning the formal algorithm for addition [1]. In Lisa's work we find this confusion again.

FIGURE 4.   Addition performed by Lisa.

Interviewer:

Okay, I'll give you one more. 2.23 plus 3.33 all in base 4.

Lisa:

Okay, okay, 6, 4 goes into 6 once with 2 remainder, I hope this is right (pause). Oh well, okay, and 5, 7, here 3 remainder, carry the 4, oh I need to be, mm, no, because I'm still in base 4, I don't want my answer in base 10, okay, so 5 plus 1 is 6, so 4 goes into 6 once with 2 remainder.

In the first addition $3 + 3$ performed in base 4 we get 12, 2 to be written down and 1 to be carried. Lisa performs addition correctly, but she writes down the 1 and carries the 2 to the next column. Therefore in her next column the addition to be performed is $3 + 2 + 2$(the carry) , which sums up to 13 in base 4. In this case Lisa manages to write down 3 and to carry 1 correctly. In the next step, there is again a reversal of digit roles and as a result 21 is written where 12 is expected.

Lisa is inconsistent in her decisions as what to write and what to carry. Looking at four other exercises done by her, we didn't find any rule for her decision making. Carrying the "wrong" digit was observed by Ashlock [1] p. 205 among frequent error computational patterns with elementary school students. Ashlock [1], p. 23, explains that difficulty children have with computational algorithms is caused by being taught computational algorithms before having adequate understanding of multidigit numerals. Lisa's interpretation of multidigit numerals, 15 years after her elementary school experiences, is still inadequate.

As discussed above, successful construction of the process of addition in bases other than ten was done by coordinating the familiar addition algorithm with a set of new basic facts. For Lisa and Mary, constructing processes for basic facts didn't seem to be a problem. One way or another the basic facts for different bases were generated accurately. The problem seemed to be with an attempt to recall the details of the familiar addition algorithm for base ten. We would like to suggest that the problem in Mary's and Lisa's case may be with *de-automization* of the process. Interiorized processes may be *automated* over a period of time. This happens when a process is repeated frequently until it can be performed mechanically and almost reflexively, without explicit awareness of the components. When an automatic process has to be reconstructed to include a wider variety of options, it has to be *de-automated* first, which means that all the components of the process have to be brought out explicitly and in detail. De-automization may be difficult, especially if the process was interiorized as a sequence of routine steps to follow without a global connection among them. It may be the case that what was interiorized as a process of addition algorithm for Lisa and Mary were procedural performances with only partial procedural understanding. An attempt to de-automate their processes reveals points of weak understanding in their well established automatic computations.

Further research is needed to study the mechanisms by which an interiorized process becomes automated and what is involved in de-automization; what makes de-automization hard for some people in some situations and easy for others. We conjecture that the difference between an "expert" and an "expert-teacher" is related to de-automization abilities.

Biggest available digit pattern in addition.

In the following excerpt Laura performs addition of 2.34 and 1.21 in base five. Whenever the sum of the digits exceeds 4, she writes down the 0, counts the number of units above 4 and carries it to the next column. Using this strategy and getting the sum of 5, that is, 10_{five}, Laura writes down 0 and carries the one to the next column, which happens to be correct. But getting the sum of 6 she writes down 0 as well and carries the 2 to the next column, since, according to Laura, 6 is two numbers above the largest one-digit number that you "can say" in this base.

Laura:

> Oh, okay. I can probably do that. This is how I would do it, are you watching? This is hard to follow (laugh). Okay, 4 and 1 is 5 but you can't say 5 because it's base 5 right? So it has to be a 0, so then you put 1, so then you'd have to put one extra one up here because it's 4,10 [read: four, one-zero], so that's an extra one up here. So that would be, 3 and 2 are 5, and one, 6, but you can't say 6, but (pause) I'm just trying to think how

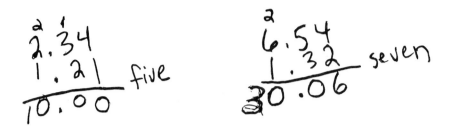

FIGURE 5.   Addition performed by Laura.

we did it the other day. Okay, now maybe this works. [counting] 4, 10 with
the 1 up there, 3 and 1 is 4 plus 2 is 6 but you can't say 6 because it's only
base 5, so it would have to be another 0 plus 2 more have to go up here, I
think. Base five, okay, 4 and 1 is 5 is 0 put the 1 up there. Four, which you
can say but plus 2 you can't say. So 0, so be it, 2, 4 plus 1 is, yea that's
what I would have to go with — 10.00.

And another example:

Interviewer:
  Um, (pause) so 6.54 added to 1.32 in base 7.
Laura:
  Five and 3 are 8 but you can't say that, so I'll put 1 up there but you can't
  say that, 5 and 2 are 7 which you can say, 5 and 3 are 8 which you can't
  say that so it's 0 with the 1 up there. 7 you can't say that, 8 you can't say
  that, so it should be 2 because 4 and 2 are 6, 5 and 3 are 8 that's · · · one
  more with the 1 up there is 7, is 8, but that's only one more too, no it's 2
  more, it's 2 more. Because 4 and 2 are 6 which you can say but 5 and 3
  are 8 which you can't say but that's 2 more over so it has to be. Then 6
  and 2 are 8, and 1–9 so that's like 3 over, so it would have to be like this.
  So 4 and 2 are 6 and then 5 and 3 are 8 and you can't say 7 and you can't
  say 8 so that's 2. So you put 2 up there, 6 and 2 are 8 and 1 are 9 but
  you can't say 9, and so it would have to be 7,8,9, so that's 3, yes, 1, 2, 3,
  [matching 1,2,3 to 7,8,9] so that would have to be · · · my answer is 30.06.

The same strategy that was used above working in base 5 appears again when
working in base seven. For example when the sum is 9, Laura writes down the
0, and then by counting and matching 1,2,3 to 7,8,9 respectively, she finds out
that 9 is 3 units above the 6, which is the largest possible one-digit number in
base seven, therefore, she carries 3 to the next column. It is obvious that using
Laura's algorithm one would get a zero in each column that involved the carry.
The appearance of an unreasonable number of zeros in Laura's answers didn't
seem to catch her attention.

A slightly different variation of the same idea appears in Sarah's work. When regrouping is needed, Sarah always writes down the largest digit possible in the given base and carries the rest to the next column. For example, if the sum in base six is 7, Sarah would write down 5 and carry the remaining 2 to the next column.

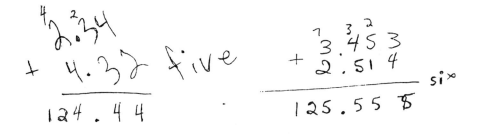

FIGURE 6.   Addition performed by Sarah.

Interviewer:

I give you 2.34 plus 4.32 all in base 5, how would you add them?

Sarah:

Okay, how I add them together. Okay, so I would look at the 1's place and go there's 6, you can't have 6, I can just have 4, so I put 4 down, and carry 2 over to the next place. Okay, now I've got 3, 6, 8, I can only have 4 again, so I'll put 4 down and carry 4 more over. Okay, so you got 4 and 6, 10, I can only have 4, so (pause) I put 4 down and I've got 6 left over, so that would be 2 in the next place and then I've got 1 group of 4 left over, so that would be 1 in the next place over.

Interviewer:

Okay. Um, how about 3.453 plus 2.514, all in base 6.

Sarah:

Okay, and we're adding again. Okay, so there's 6, you can only have 5 here, so I have 2 in each to carry over, and you got 12 here, so that makes it, 12, I think I would take 2 over to this place and then, okay, okay, just a minute here. It's mixed up. Okay, so that's 7 here because we've got 5, and we got 2, okay, I think I've been doing this wrong. Okay, um, I have to think for a minute now. (Pause) I'm going to have to rearrange these numbers, for one thing, okay? (Pause) Okay, so 7 is in the third place over from the decimal, okay, okay but I can only have 5, so we've got 2 left over that I bring over to the second place and I can only have 5, so I've got (pause) 3 I carry over here, okay, so I have to carry 7 over, okay so in the one's place there would be 5 and I've got 7 left over, so that would be 2 here and 1 here, so it would be 1–125.555.

Sara's strategy is reminiscent of arranging discs on pegs, ignoring the idea of place value. Each peg can hold 5 disks only. Sara seems not to apply regrouping

rules, but to move the extra discs/units to the next peg/column on the left.
That's how 6 and 7 got represented in base 6 as 15 and 25 respectively.

For Laura and Sarah the source of error in reconstructing addition algorithm
seem to be with generating a counting sequence for unfamiliar bases. When
such a sequence has to proceed beyond single digit numbers, Laura's counting
sequence for base seven appears to be $\cdots$ 5, 6, 10, 20, 30; and Sarah's counting
sequence for base six was $\cdots$ 5, 15, 25, $\cdots$. Such discrepancy may be explained,
again, as a difficulty in de-automatization of the counting process, that reveals an
incomplete understanding of place value numeration.

Conversion Tasks.

As mentioned earlier, only 10 out of 20 students performed correctly on problems of conversion. In what follows we discuss students' strategies and possible
constructions.

Conversion — expanded notation.

Ann draws explicit analogies in translating her base ten understanding to the
understanding of base 5.

Interviewer:

Could you convert a number like 3.24 in base 5 into base 10?

Ann:

Well that's interesting. Um, I think I could. I could try. Do you want me
to do it?

Interviewer:

Please do it. Yeah.

Ann:

Okay. Well my 3 is still the ones, so I still have the 3 and this would be,
it's interesting, so I'd have to think about what it means in base 10 first.
So, in base ten 3.24 would mean the 2 would be the 10ths so what would
that mean, fifths? Yeah, I guess it would be 5ths, so it would be, I'm not
sure how I would convert it though. So this would be 2/5ths and 4/25ths,
but I'm not sure how I would write that. I would like if I could write it in
fractions, I would say it's 3 plus 2/5ths plus 4/25ths.

In her interpretation Ann refers to the place values of digits in base ten — "2
would be the 10ths" — and then translates these values to the analog in base
five — "I guess it would be 5ths." We see that Ann starts with a familiar object
of a number 3.24_{ten} and probably de-encapsulates this object to a process of
$3 \times 1 + 2 \times 1/10 + 4 \times 1/100$. Later, she substitutes the appropriate powers of
five instead of powers of ten to construct the process of representation in base 5
— "it's 3 plus 2/5ths plus 4/25ths."

Nine out of ten students, who performed the conversion correctly, used methods similar to Ann's. Linda's conversion below was unique in this sample.

Conversion — via common fraction in another base.

Interviewer:

How about a number like 1.23 in base 4?

Linda:

Okay, um that would be 2, uh, okay that would be 23/100 [read: twenty three over one hundred] meaning three groups of 1, two groups of 4, divided by one group of 16, zero 4's and zero 1's and that would be one unit, one point two three over 100th.

Interviewer:

Okay, could you explain this to me again? I'm not quite sure what?

Linda:

Not quite sure what I mean?

Interviewer:

Yeah.

Linda:

Well what I mean is that um, well I, yeah I'm just doing this in base 10 again, I guess, I'm just using this decimal point to divide each number to the right of the decimal point over um, like 4, 4^2 each, uh, yeah, 2 groups of, hmm, well 23 over 100 over 100th, okay, but that stands, that's not 23, that is 2,3 [read: two–three] in base 4 which is 2 groups of 4 plus 3 units, okay? Does that make sense?

Interviewer:

So, if we translate - you're saying this is 2 groups of 4 plus a group of 3?

Linda:

Three, yeah, plus a group of 3 units, 3 one's divided by, okay, over 100, now the 1 in this case stands for 1 group of 4^2 okay, because of the same thing as in base 10, like you've got the 1's, you've got the 4's, you've got the four squares, these are all your columns, your place value columns, four cubed. Okay? So I'm treating it the same way as I do the base ten system, I'm always in decimal points, we always divide it over um $\cdots$.

Linda's unconventional strategy confused the interviewer, but turned out to be correct. Unlike Ann and many others, who assign place value to every digit and work with expanded notation of the number, Linda treats the fractional part as one conceptual entity. Her conversion doesn't go directly to base ten, but passes through "common fraction in base four." Indeed, the number $.23_{\text{ten}}$ is read "twenty three hundredths," which means 23/100. Therefore, for Linda, the fractional part of the number .23 in base four is de-encapsulated to a process 23/100, where both 23 and 100 are base four numbers. In her next step the integer numbers in the numerator and the denominator are both converted to base ten, and $23_{\text{four}}/100_{\text{four}}$ is interpreted as a fraction 11/16.

Conversion — moving the "decimal" point error pattern.

In a new situation, where no formal algorithm for symbolic manipulation was provided, Joan tried to use familiar problem solving strategies. One such strategy was applied when performing division by a decimal fraction: the decimal point "is moved" in order to work with integers. Using this technique when converting from non-decimals to base ten, Joan moved the point to get the integer number, converted the integer number to base ten and then relocated the decimal point in the answer.

Interviewer:

Would you be able to convert something like 3.25 base 6 into base 10?

Joan:

Okay, um this is 3 single ones, so 3 single 6's, so that's still 3, point 2, okay that's 25, ?? (pause), I would probably, I don't know, I probably would have to move out to the decimal point — I can do that. So represent · · · I'm not sure if I'm doing that right. (pause)

Interviewer:

Alright, what have you done so far?

Joan:

Well, I tried to move the decimal point because I'm more familiar with the 1's, 10's and the 100's than I am with 10ths and 100ths. But, in order to move a decimal point, and I can't just move a decimal point for the sheer heck of it, · · ·

Interviewer:

Okay, so what do you have to do when you move a decimal point?

Joan:

Well I'm multiplying. Like, if I'm going to move a decimal point that would be mult · · · like if I have 3.2 and I want to move this decimal point to make it a whole number, then I would multiply it by 10's and my decimal point would go, so if I'm moving this decimal point two places I multiply by 100, but I'm not familiar enough with this to know that if I once have an answer I can divide by 100 and adjust the decimal points. (pause) Whaa! not a nice question.

In order to convert 3.25_{six} to base ten, Joan moved the "decimal" point two places to the right, then she converted the number 325_{six} to base ten. She got 125 and "adjusted" the decimal point moving it two places to the left to get the result of 1.25. She wasn't sure in what she did, but it was "the only way (she) could think of."

Joan was asked to reconstruct a familiar process of converting integers from base six to base ten to include mixed numbers. Instead, she preferred to avoid reconstruction by changing one situation to another, in which application of her original process is possible. In some other cases this may be a useful heuristic. Here, Joan didn't realize that by relocating the "decimal" point two places to the right she was actually multiplying by 100_{six}, that is, by 36. Therefore the ratio between the mathematically correct answer, $3\frac{17}{36}$ or $\frac{125}{36}$, and Joan's answer is 100 : 36.

Conversion — reading bug error pattern.

Michelle applies a strategy that was referred to by Zazkis and Khoury, [26] as a "reading bug." Rather than assigning place value to each digit in the fractional part of the number, she treats the fractional part as one entity and assigns to it the "place value" of the smallest unit involved. That's how 0.12_{four} is interpreted as 12/16 (in base ten). The name "reading bug" refers to the reading rules for decimal fractions: saying "twenty-three hundredths" brings to mind 23/100, rather than 2/10 and 3/100. The equality between the two expressions was misinterpreted when applied to other-than-ten bases.

Interviewer:

How about converting 3.12 base 4 into base 10?

Michelle:

Umm, yeah sort of, the fractions kind of stump me. Okay, so this is basically 12/100ths, so oh, it isn't 12/100ths, it's 12/16ths, it's 3 units, so it's 3 ones, it's 12/16ths.

Interviewer:

How about something like 4.34 in base 5 convert that to base 10.

Michelle:

See, I think that I could see this better if I actually um, okay this is groups, this is 34/25, no, that can't be right, because it's more than a whole. Yeah, you can't have that, that's wrong.

Interviewer:

(Laugh) And why is that wrong?

Michelle:

Because 34 is too big, so this is actually 5.34 minus 25 which is 5.9 or 09, 9 · · · 5.9 or 5.09? I'm not sure. Hmmm. Instead of 34/25ths it's 9/25ths so it's 09.

Interviewer:

And the number 6.54 uh base 7, could you convert that into base 10?

Michelle:

So then that's 7.05.

Interviewer:

Alright, and maybe you could just, for the record, explain to the tape how you're doing these questions.

Michelle:

Well, I'm um, well because, because my placement value is to the right of the decimal, or two placements over, I'm taking my base number and multiplying it by myself, by itself because I'm not in my units, I'm in my, whatever you call that, my cubic (laugh). Anyway, I'm multiplying 7 times

7, the base times the base to give me my, my bottom number, so instead of 100ths I have 1/49ths and then I'm just subtracting the difference. And then, when I'm subtracting the difference that's giving me one whole unit, so I'm adding that unit on to my units and then I'm leaving the difference, and I'm putting the 0 because I'm, my subtraction is over 49, which is equivalent to my 100ths because it's over my 49ths, that means that I have to put a 0 because it was over 1/7th.

Similarly to Linda's construction, Michele de-encapsulates 0.12 to a process of 12/100 — "this is basically 12/100ths." Unlike Linda, when such de-encapsulation has to address base four, only denominator was treated as base four number — "oh, it isn't 12/100ths, it's 12/16ths."

In the second and third example, Michelle revised the "reading bug" for cases where the numerator appears to be bigger than the denominator. In the last conversion discussed, 6.54_{seven} is equal, according to Michelle's strategy, to $6\frac{54}{49}$ Having an improper fraction in the fractional part seems an improper representation to Michelle, so she applies familiar processes for mixed numbers to change her answer to $7\frac{5}{49}$. Moving in the next step from the common to the "decimal" fraction representation this number becomes 7.05.

Conversion — common fraction substitution error pattern.

Marina uses a pattern that was named by Zazkis and Khoury [26] "common fraction substitution." She interprets the place values in the fractional part of non-decimal representation as 1/4 and 1/40 or 1/6 and 1/60 for bases four and six respectively.

Interviewer:

Alright, where did you get the numbers 3/4ths and 2/40's from?

Marina:

Well because the 4 is the reverse of this one, because the place value goes like this, the place value goes 1, like in 10's, it goes 1's, 10's, 100's or 10 squared and then here on the other side of the place value it would go, um 1/10th, 1/100th, so that's why I was thinking of that but I get confused when I try and convert it into 4ths.

Interviewer:

And if you had a number like uh .452 in base 6? What is your first reaction to that?

Marina:

That would be, that would be also the reverse, that would be 4/6,5/60ths and 2/600ths.

From Marina's interview it seems the case that the object of the mixed number was encapsulated to a process of its expanded notation, but then the place values for bases other than ten were not constructed properly. Even though Marina mentions exponential notation in naming place values for base ten — "it goes

1's, 10's, 100's or 10 squared" — she doesn't seem to make use of it when constructing place values for other bases. She correctly interprets the first place value to the right of the "decimal" point to be 1/6, but in constructing place values, she seems to use an analogy with a feature base ten, in which one place value is different from a neighbor just by one zero.

The following excerpt is from an interview with Celia, who started with an approach similar to Marina's, but proceeded further.

Interviewer:

Okay, how about if I were to give you a number like 4.52 in base 6? What would that mean to you?

Celia:

In base 6? I've never looked at in decimals before. Well, if it was, if it was base 10, we were talking about this in class and how you represent the right side of the decimal $\cdots$. These are ones then what are these? One tenth, one hundredth, so if you were doing that in base 6, 1/6th, 1/60th ? 4 times 1, and so you're asking me, what was the question about this number?

Interviewer:

What does it represent?

Celia:

What does it represent. (Pause) So that would be 4 times 1 on this side, and that would be 5 times 1/6 and that would be 2 plus 2 times 1/60, doesn't make sense. (Pause) I've never even looked at it in base six before.

Interviewer:

[referring to what Celia writes on paper] So is an answer you have 4 times 1 plus 5 times 1/6 plus 2 times 1/60? $[4 \times 1 + 5 \times (1/6) + 2 \times (1/60)]$

Celia:

I guess, but I don't know, that doesn't — I don't whether that's uh, whether that really makes sense. (Pause) One thing that I'm confused about is because we have 2 digits here and only 1 digit here, 3 digits here and only 2 here and yet we're in base 6. So maybe that's okay, I don't know. (Pause)

Interviewer:

Where did you get the 1/6 and the 1/60 from?

Celia:

Well, just because I'm thinking of um, just looking back to this, what we were looking at before. Um, (pause) 36 as opposed to, because we're talking in square roots not 10's, 6, 1/36? I don't know.

Interviewer:

So, now you've changed your answer from $5 \times (1/6) + 2 \times (1/60)$ to 5 times 1/6 plus 2 times 1/36.

Celia:

Because we're talking about square, squares of numbers. I was, when I had written 60, I was still thinking that base 10 is not correct, talking about 6

groups of 6, but I'm still confused about the uh, there's only one digit and
there's 2 here and there's 2 digits and there's 3 here, so I don't know.

Interviewer:

Do you think the digits, the number of digits could be different?

Celia:

Well, I don't think they should be different, $\cdots$. And yet I have them as
different. Like maybe, going back to this other one, this should be 1/60
and 1/360 for that, for the $\cdots$. I don't know. Can't think of something
else.

Similarly to Marina, Celia initially interprets the place value of "hundredths"
in base six as 1/60. Then she changes her mind to 1/36, explaining that "we're
talking about square, squares of numbers." With this change Celia faces disequilibration, since the number of digits in the denominator of 1/36 is different from
the number of digits in the denominator of 1/100. To resolve this disequilibration Celia changes her answer to 1/360. As we see, Celia wasn't sure about her
answer, but "couldn't think of something else."

We would like to note here that common fraction substitution pattern leads
to correct answers in cases of only one digit in the fractional part of the number.
That's probably why Deborah managed to convert correctly the number 12.2 in
base 4 to base ten. When asked to convert 12.21 in base 4, Deborah wrote her
answer as $6\frac{21}{4}$. As a response to interviewer's question "where did you get 21/4
from" Deborah corrects her answer and changes 21/4 to 21/40. Her pattern
seems to be a combination of reading bug and common fraction substitution
strategy.

Deborah:

2 and 4 that's 6 point $\cdots$ and that's 21 over 40 because otherwise it would
be 1/10th or 1/100th, you add a 0 because you're going further down, I
don't know if it works in this. Yeah, it would be 6 plus 21 over 40 (pause)

Deborah's process of generating place values in base ten — "you add a 0 because you're going further down" — has been wrongly generalized. It is possible
that the reason for the effect of "adding zeros," which is , in fact, multiplication
by the base of ten, wasn't clear to Deborah.

Conversion — "tens-or-tenths" confusion error pattern.

David seemed confused in the beginning by the task of conversion. Therefore
the interviewer referred to base ten.

Interviewer:

Let's look at 632.15 in base 10. What does each digit represent?

David:

Yeah, I just drew out any number and then what I did was I wrote under-
neath what each column was worth, so your 2 is your units column and
your 3 is your 10's column and your 6 is your hundred's column and on
the other side of the decimal your 10's column is the first column and your

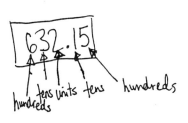

FIGURE 7.   Conversion performed by David.

100s column is the second column. So if you look at 0.4 on base 6 system, 4 takes the place of the 10's, it's the 10's place so what you need to do there, if you follow the same strategy is, you take 4 groups of 6 is 24, so because 0 is just a place value holder, you still have the 0 there, so you have 0.24 in base 10, so 0.4 in base 6 is 0.24 in base 10.?

Interviewer:

Okay, um I'd like to try one more example, at least. So, let's try .32 in base 4. How would you convert that to base 10?

David:

Okay so your 3 is in the 10's column so that would like 3 groups of 4 that's 12, and your 2 is in your 100's column so that's like 2 groups of 42 so that's 42, 16 times 2 is 32 so then you add those 2 together, that's 44, so in the base 10 system that would be 0.44. Zero point 32 in base 4 would be 0.44 in base 10. I guess what you have to do is make sure that you know what, know what column you're in and what procedure to use because if you can't remember what place value you're holding then you're never going to be able to figure out what to do with that number to make the conversion and that's where my problem was coming from before because I was getting mixed up as to what each um, each column represented. Because when you're on the other side of the decimal the um pattern changes, you're reading from left to right instead of from right to left, yeah.

Our first reading of the interview looked like an oversight by the transcriber. Listening to the tape didn't help — we couldn't confirm the difference between "tens" and "tenths" in a fluent narrative. Only looking at the notes (Fig 7) made by David during the interview confirmed the misconception and the confusion: David labeled the place values as "ones," "tens" and "hundreds" to the left of the decimal point, but he also labeled "tens" and "hundreds" to the right of the decimal point. Such labeling, with the choice of appropriate values for each one of the base systems, was implemented in all his calculations. That's how 0.4_{six} was interpreted as 0.24_{ten}, since $4 \times 6 = 24$. That's how 0.32_{four} was calculated by $3 \times 4 + 2 \times 16 = 44$ and converted to 0.44_{ten}.

Similarly to other participants, like Ann, Marina, Celia or Deborah, David de-encapsulates the object of a number to a process by assigning place values to its digits, as if it were base ten representation, and then he reconstructs the place values taking into consideration the base in question. Unlike the others, David's assignment of decimal place values is incorrect, but being carried over precisely, following the "same strategy," in assigning place values in other bases.

Fuson [13, 14] and Resnick et al. [20] discussed the problems that inconsistent English number words add to the learning of place value. David's confusion between "tens" and "tenths," between "hundreds" and "hundredths" can also be blamed on the number words in the English language. However, this would be an easy way out to explain the innumeracy of a person, who is only 6 months away from entering the classroom and taking responsibility for the mathematical education of elementary school children.

Conversion: Mirror in the "decimal" point error pattern.

In the very beginning Mary attempted to use the "separation and gluing strategy" [26]. In this strategy the integer and the fractional part of the number are treated separately as two integer numbers, they are converted separately and then the results are "glued" together with a decimal point in between. It seemed like for Mary the object of a rational mixed number was de-encapsulated to a process of writing two integers side by side with a "point" in between. When asked to explain this strategy, Mary changed her mind claiming that "on the right side the place values have to go the other way around." Therefore Mary converted 4.215 in base six in the following way: $4.215_{\text{six}} = 4 + 2 \times 1 + 1 \times 6 + 5 \times 36 = 4.188$.

Mary:

> Thinking back to 4.215, so if I took that, that's a 4, this stays a 2, but then this is 1 group of 6, 1 group of 6, this group's 6, so that's plus 6 and this is 5 times $\cdots$, 36 times 5 which is what? 180 that which is 180 plus 180, so that's, so then the answer would be 4.188, oh, now this is getting confusing.

It may be still the case, especially after a closer look at the algorithm that Mary used for addition, that for her the object of the rational number is constructed by "gluing" two integers, while the place values of the digits of the integer on the right are mirror images of the place values of the digits of the integer on the left.

Conversion — what comes after assigning place values.

It seemed to us that what was left in the conversion task after assigning place values to digits, was just a matter of simple arithmetic. Marina and Jill convinced us that the issue isn't that simple. Not that they failed in "simple arithmetic," they just had difficulty to decide what arithmetic had to take place.

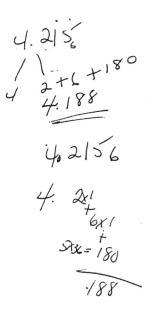

FIGURE 8. Conversion performed by Mary.

Marina felt stuck after assigning place values to the digits of number 1.32_{four} and didn't know how to proceed with the task.

Interviewer:

Alright, and what have you written here? You've written it's going to be 1 and then you have 3/4ths and 2/40's. And what do you do with those? (pause) Would you add those?

Marina:

No, I wouldn't add them together, but how would I figure it out you mean? I don't know if I can or not without using, I probably couldn't without using manipulatives instead of writing it out. (pause) Well, I would have to just, I would have to count them out.

Interviewer:

And how would you count them?

Marina:

I wouldn't be able to convert it into base 10 unless I had some kind of manipulatives because of the decimal point. I would have to see it broken down. I would have to see it visually broken down.

Jill assigned place values correctly, but didn't look at the sum of the products. Instead, she tried to estimate the number:

Interviewer:

How would you convert to base ten number 6.42 in base 7?

Jill:

> This will be units, 1/7th, 49ths, 6 units, 4/7ths and 2/49ths, yeah, 2,
> yeah, so then 2/49 which is approximately, I don't know, I guess I can't
> approximate, but that's sort of the same as 4/100 close to 2/50ths, I guess
> would be .04. I'm rounding is what I'm doing now, because I don't know
> how to calculate, 4/7ths is more than, I don't know what that is in a
> decimal, I can work it out. Um, this shows you what you don't know,
> anyway. (pause)

It may be the case that the confusion that Jill faces is in converting simple fractions to decimal fractions. We find both Marina and Jill at the stage
where they have de-encapsulated the object of a number to a process, have reconstructed the process taking into account the new base in question, but their
process didn't go beyond assigning place values to the digits of multidigit number. This process was insufficient to complete, either correctly or incorrectly, the
task of conversion. The issue of the process resulting from de-encapsulation will
be discussed further in the next section.

INTERPRETATIONS AND DISCUSSION

According to our theoretical perspective, students' ability to perform addition/
subtraction with non-decimals indicated that they treated these numbers as *objects*, since they were able to perform *actions* on them. The conversion task
requires students to move from an object back to a *process*, which is done, according to the theory, by *de-encapsulating* the object. The objects (rational
numbers) that had to be de-encapsulated by the interviewees were constructed
a long time ago and probably over a long period of time. How difficult or how
natural is the de-encapsulation in this case? Dubinsky [9] conjectured that "an
epistemological obstacle occurs when an action, process or object conception of
some mathematical topic existed for some time and was very useful in dealing
with many problem situations, but now is presented with problems that it cannot
handle." Our findings definitely support this conjecture and provide an example
of partial or incorrect place value conceptions that served our subjects faithfully
throughout their college years, but caused an obstacle when used to interpret
non-decimals. Our results confirm that de-encapsulation takes the student back
to the process which was encapsulated in order to construct the object in the first
place. This process may not be sufficient to deal with problems on a higher level
of sophistication and needs to be reconstructed. This reconstruction appeared
to be very difficult for some students and was a major source of errors.

Teaching Computational Algorithms.

There are different perspectives in mathematics education regarding the teaching of computational algorithms. Fuson [13, 14] argues that multidigit addition
and subtraction create a rich environment for the acquisition of place value
concepts. Dienes' blocks are the most commonly used materials in teaching

multidigit addition/subtraction and the effectiveness of instruction using these manipulatives has been strongly advocated (e.g., [**5**, **6**, **19**, **24**, **25**]). Other researchers (e.g., [**4**, **16**]) illustrate students' extensions and connections in procedures they invent when solving problems of multidigit computation prior to being introduced to the formal computational algorithms. Their claim is that making connections explicitly for the students by introducing them to the formal algorithms and or by using specially designed materials prevents students' attention from being distributed across the range of possible connections.

Our findings in this research show that a significant number of students, 6 in a sample of 20, were able to perform addition and subtraction with non-decimals correctly, but were not able to perform the conversion. Since assigning place values to the digits is a substantial portion of the conversion task, these students' place value concepts were probably not sufficiently developed. However, their place value concepts didn't interfere with reconstructing the standard algorithm for multidigit addition and subtraction and adapting those algorithms to computations in various bases. This finding suggests that knowing the place values of specific digits isn't necessary for addition/subtraction tasks. Four and three added together make seven, regardless of whether 4 and 3 represent oranges or apples, tens or hundreds, tenths or thousandths. The result of the same addition in base five is represented by 12_{five} and when performed as part of a bigger calculation will always mean "write down 2 and carry the 1 to the column on the left," regardless of the place value of the columns discussed. We would argue that multidigit addition and subtraction require only a partial understanding of place value, and we doubt whether a fuller understanding may be achieved through multidigit computational tasks.

An additional observation in this study is that addition with non-decimals caused "more troubles" than subtraction. In making this observation we're reporting not just the ratio of correct performances in one versus another , but our overall impression taking into account the uncertain tone in student's talking aloud and the amount of time it took to perform the computation. It may be a very specific feature of our sample. It may also be the case because addition problems were presented to our interviewees before the subtraction problems. But it could be the case that the formal subtraction algorithm is less demanding or easier to generalize. Since traditionally addition is taught before subtraction, we see the need for examining this observation in more detail.

Constructing non-decimals.

Ann and Linda demonstrate two different but mathematically correct interpretations of non-decimal representation of a rational number. In both cases the object of non-decimal is de-encapsulated to a process. For Ann the object of 0.23 is de-encapsulated to $2/10 + 3/100$. For Linda the object 0.23 is de-encapsulated to $23/100$. Using the base ten analogy, both Ann and Linda coordinate this process with the process of representing the *fundamental sequence* for a given base. In what follows we discuss the concept of the "fundamental sequence."

In base ten representation the fundamental sequence is a geometric sequence of powers of ten. The construction of this sequence starts as early as children learn to assign to the digits of multidigit numbers values of "tens and ones." This sequence is reconstructed and expanded later to become "hundreds, tens and ones" and "thousands, hundreds, tens and ones" and so on. When the learners are introduced to the idea of decimal fractions, this fundamental sequence is reconstructed to include "tenths, hundredths, thousandths, etc.". It is later understood that this sequence is infinite in both directions, that is,

$$\cdots, \ 1000, \ 100, \ 10, \ 1, \ 1/10, \ 1/100, \ 1/1000, \ \cdots \quad \text{or}$$

$$\cdots 10^3, \ 10^2, \ 10^1, \ 10^0, \ 10^{-1}, \ 10^{-2}, \ 10^{-3}, \ \cdots.$$

Assigning place values to the digits of multidigit mixed decimal numbers may be no more than matching the digits to the elements of this sequence. In order to create a fundamental sequence of another base, the decimal sequence has to be de-encapsulated to a process. In this process the learner has to be able not just to name the elements of the sequence, but to create the next element in line and explain the relationship between the elements. If this process means to the learner "taking consecutive powers of the base of ten" or "multiplying by ten to create the element to the left and dividing by ten to create the element to the right of the given one," it can be easily reconstructed and generalized to the process of fundamental sequence in any given base. In an example of base five, the fundamental sequence is

$$\cdots 125, \ 25, \ 5, \ 1, \ 1/5, \ 1/25, \ \cdots \quad \text{or}$$

$$\cdots 5^3, \ 5^2, \ 5^1, \ 5^0, \ 5^{-1}, \ 5^{-2}, \ 5^{-3}, \ \cdots.$$

As we show in the next section, many of the error patterns in converting non-decimal representations of fractions to base ten were caused by generating inappropriate fundamental sequences.

To proceed with construction of non-decimals, coordinating Ann's process with the fundamental sequence process leads to expressing 0.23_{five} as a sum $2/5 + 3/25$. Coordinating Linda's process leads to expressing 0.23_{five} as $23_{\text{five}}/100_{\text{five}}$, which is $13/25$. We summarize those constructions in Figure 9.

Constructing non-decimals — What can go wrong.

Many of the error patterns that were observed in this study can be explained as a defect in one of the transitions described in Figure 9. The reading bug pattern for example is an inadequate coordination of Linda's process with a fundamental sequence. Instead of interpreting 0.23_{five} as $23_{\text{five}}/100_{\text{five}}$, students using the reading bug interpreted this number as $23_{\text{ten}}/100_{\text{five}}$.

The common fraction substitution error pattern results from constructing the fundamental sequence of place values in base five as $1/5, 1/50, 1/500$ etc.. This happens when a process of $1/10, 1/100, 1/1000$ is constructed by "adding zeros," as Marina stated, rather than dividing by ten or multiplying the denominator by 10. This also may happen when the learner acknowledges multiplication by

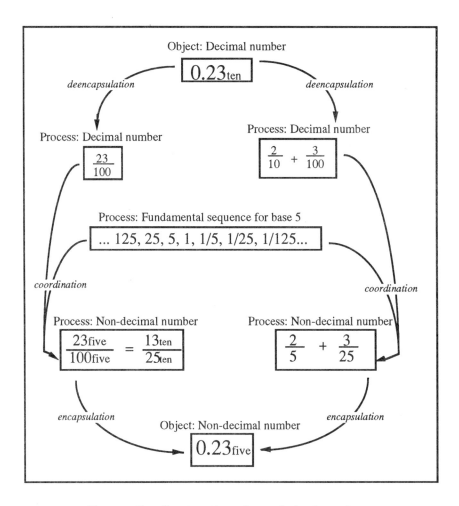

FIGURE 9.   Construction of non-decimal number.

ten, but doesn't focus on the essence of ten as a base and treats it as a constant value to generate sequences for other bases.

In the responses of David and Mary we see references to "the other part of the decimal," "the opposite side," "what is here reversed there," "reading from left to right instead of from right to left." What is captured in these responses is the process of "flipping over," of generating the sequence for the fractional part as some sort of reflection of the integer part. For David the fundamental base ten sequence was 100, 10, 1, 10, 100, $\cdots$, so this is not surprising that this process was reconstructed to base five fundamental sequence of $\cdots$, 25, 5, 1, 5, 25, $\cdots$. We note that this sequence is symmetric and the "one's" place is the "line of symmetry." Mary explicitly treated the fractional part of a number as a mirror image of the integer part. Her addition and subtraction were performed separately on each part, going left to right in the fractional part. Her fundamental sequence was also symmetric, but differed from David's in the line of symmetry. The sequence generated by Mary was $\cdots$, 25, 5, 1, 1, 5, 25, $\cdots$, where the "line of symmetry is in the 'decimal' point." Although the decimal point marks the transition from integers to fractions, the fundamental relationship between the columns is not changed, for all positional system representations. This basic relationship was not discovered by Mary and some other of our interviewees.

The understanding of a multidigit structure requires more than the understanding of place value [1]. It requires an understanding that the multidigit number is a sum of the products of its digits , so called "face values," by place values. Even though assigning place values is the essence of the conversion task, it is insufficient to solve the problem. It may be the case that for Marina, for example, the object of 0.23 is de-encapsulated as "two tenths and three hundredths," where "and" means concatenation, writing values sided by side, but not addition. This may explain why Marina couldn't proceed in the task of conversion beyond assigning place values.

Evidently, students' constructions of the fundamental sequences in other bases, their de-encapsulation of fractional decimal numbers, as well as their de-automization of addition algorithm, suggest where they may have deficiencies in their understanding of the base ten numeration system.

CONCLUSION

We were able to analyze most of the data in terms of constructing actions, processes and objects, mainly de-encapsulating objects into processes and coordinating processes into new processes. This suggests that the action-process-object theoretical framework is a useful language to describe and analyze students' constructions in the domain of non-standard representations of fractional rational numbers. The use of this theoretical perspective suggests generalities regarding construction of mathematical knowledge, applicable to learning "advanced" collegiate mathematics as well as to learning "not-so-advanced" collegiate mathematics.

The results suggest that preservice teachers' constructions of place value number system are fragile and incomplete. Teachers' partial understanding doesn't interfere with their correct algorithmic performance, but may result in incomplete constructions in the minds of young learners. We believe with Steffe [22] and many others that the improvement of mathematics education in our schools starts with the improvement of mathematical knowledge of teachers. Place value is a basis of positional system arithmetic. We believe that this analysis of teachers' knowledge in this domain in this study highlights possible pitfalls and conceptual difficulties.

According to Ashlock [1], the children have difficulties with computational algorithms because they do not have an adequate understanding of multidigit numerals at the time they are introduced to the algorithms. When such children become teachers, many of them still hold inadequate understanding of multidigit numerals.

Dienes [**5, 6**] advocated the use of multibase arithmetic to reinforce under-standing of the base ten positional system. We find renewed interest in this topic at the college level, for education of preservice mathematics teachers. Hungerford [**15**] suggests and strongly encourages a new instructional approach for teaching a "new arithmetic," since it "seems to improve students' understanding of the mathematics involved. On the other hand, Freudenthal [**12**], arguing with "innovators," who "like to do structures on other bases," claimed that "if compared with mathematics resulting from pondering more profoundly the subject matter and its relations to reality, unorthodox positional systems are a mere joke." Even so, Freudental didn't exclude other-than-ten bases for "remedial use," and actually stated that "it is a good didactics to motivate pupils by jokes, and an unorthodox positional system may even be a good joke."

Regardless of what one may call the learning/teaching experiment with other-than-ten bases — be it a powerful tool or a "good joke" — we appreciate its value as a topic leading to rich and deep mathematical investigations, and as a topic assisting in acquiring deeper understanding of place value numeration and multidigit structures. We suggest that "non-decimals" become an integral part of this topic.

REFERENCES

1. R. B. Ashlock, *Error Patterns in Computation: A Semi-programmed Approach, fifth ed.*, Merril, Columbus, OH, 1990.

2. T. Ayers, G. Davis, E. Dubinsky, and P. Lewin, *Computer experiences in learning composition of functions*, Journal for Research in Mathematics Education **19:3** (1988), 246–259.

3. D. Breidenbach, E. Dubinsky, J. Hawks, and D. Nichols, *Development of the process concept of function*, Educational Studies in Mathematics (1991), 247–285.

4. T.P. Carpenter and J. M. Moser, *The Acquisition of addition and subtraction concepts in grades one through three*, Journal for Research in Mathematics Education **15** (1984), 179–202.

5. Z.P. Dienes and M.A. Jeeves, *Thinking in structures*, Hutchinson, London, 1965.

6. Z.P. Dienes, *Building up mathematics*, Humanities Press, New York, 1960.

7. Ed Dubinsky, *Constructive aspects of reflective abstraction in advanced mathematics*. Epistemological Foundations of Mathematical Experience (L.P. Steffe, eds.), Springer-Verlag, New York, 1991.

8. Ed Dubinsky, *Reflective Abstraction in Advanced Mathematical thinking*. Advanced Mathematical thinking (D. Tall, eds.), Kluwer Academic Publishers, Boston, 1991.

9. Ed Dubinsky, *A Theoretical Perspective for Research in Learning Mathematics Concepts: Genetic Decomposition and Groups*. Paper presented at the Sixteenth International Conference for Psychology of Learning Mathematics (1992), Durham, New Hampshire.

10. E. Dubinsky, U. Leron, J. Dautermann, and R. Zazkis, *On learning fundamental concepts of group theory*, submitted.

11. J. Fauvel and Jeremy Gray (eds.), *The History of mathematics : a reader*, Macmillan Education in association with the Open University, Basingstoke, 1987.

12. H. Freidenthal, *Didactical Phenomenology of Mathematical Structures*, Reidel, Dordrecht, The Netherlands, 1983.

13. K.C. Fuson, *Issues in place value and multi-digit addition and subtraction learning and teaching*, Journal for Research in Mathematics Education **21** (1990), 273–280.

14. K.C. Fuson, *Conceptual structures for multiunit numbers: Implications for learning and teaching multidigit addition, subtraction, and place value*, Cognition and Instruction **7:4** (1990), 343–403.

15. T. W. Hungerford, *An experiment in teaching prospective elementary school teachers* **4:1** (1992), UME Trends.

16. C. Kamii and L. Joseph, *Teaching place value and double column addition*, Arithmetic Teacher **35:6** (1988), 48-52.

17. P. Nesher, *Towards an instructional theory: The role of student's misconceptions*, For the Learning of Mathematics **7:3** (1987), 33–40.

18. J. Piaget, *The Child's Conception of Number*, Routledge and Kegan, London, 1952.

19. L.B. Resnick, *The role of invention in the development of mathematical competence*. Developmental Models of Thinking (R.H. Kluwe and H. Spada, eds.), Academic Press, New York, 1980, pp. 213–244.

20. L.B. Resnick, P. Nesher, F. Leonard, M. Magone, S. Omason and I. Peled, *Conceptual bases of arithmetic errors: The case of decimal fraction*, Journal for Research in Mathematics Education **20** (1989), 8–27.

21. J.V. Trivett, *And So On: New Designs for Teaching Mathematics*, Detselig Enterprises Limited, Calgary, Albert, 1980, pp. 24–39.

22. L. Steffe, *On the knowledge of mathematics teachers*, Constructivist views on the teaching and Learning of Mathematics, JRME Monograph No.4, NCTM (Robert B. Davis, Carolyn A. Maher, Nel Noddings, eds.), Reston, Virginia, 1990, pp. 167–184.

23. K. VanLehn, *Mind Bugs: The Origins of Procedural Misconceptions*, MIT Press, Cambridge, MA, 1990.

24. D. Wearne and J. Hiebert, *A cognitive approach to meaningful mathematics instruction: Testing a local theory using decimal numbers*, Journal for Research in Mathematics Education **19** (1988), 371–384.

25. D. Wearne and J. Hiebert, *Cognitive changes during conceptually based instruction on decimal fractions*, Journal of Educational Psychology **81** (1989), 507–513.

26. R. Zazkis and H. Khoury, *Place value and rational number representations: Problem solving in the unfamiliar domain of non-decimals*, Focus on Learning Problems in Mathematics, vol. 15:1, 1993.

27. R. Zazkis and D. Whitkanack, *Non-decimals: fractions in bases other than ten*, International Journal of Mathematics Education in Science and Technology **24:1** (1993), 77–83.

SIMON FRASER UNIVERSITY

NORTHERN ILLINOIS UNIVERSITY

CBMS Issues in Mathematics Education
Volume 4, 1994

Twenty Questions about Research on Undergraduate Mathematics Education

LYNN ARTHUR STEEN

"A pump, not a filter" is the banner of reform in undergraduate mathematics. Almost everyone agrees that the current system is not working as well as it might. Some pipes in our educational system are virtually clogged, others leak badly, while many are disconnected from their destination. For far too many able students, undergraduate mathematics fails to inspire, to motivate, or to lead. Put simply, our educational system often fails to educate.

In many arenas of human endeavor—e.g., engineering, medicine, technology—the typical approach to a challenge of this sort is an intensive R&D effort. Basic research, leading to applications, yields necessary insight for the design of new systems that respond to the challenge. Mathematicians know this paradigm well, since our discipline is the language in which most R&D takes place. But when confronted with problems of our own making—of undergraduate mathematics education—mathematicians rarely recognize research as a valuable or natural part of the response.

Many claims are made to explain this anomaly—claims that appear to some as sound reasons, to others only as lame excuses. These conflicting claims must be addressed by both advocates and critics if research in undergraduate mathematics education is to achieve either legitimacy or utility. Some of the issues reverberate around mathematics departments either as rhetorical questions or as skeptics challenges. These questions define the challenge and frame the debate.

1. *What are the goals of educational research?* What, indeed, are the goals of education? Is the purpose of educational research to understand education or to improve it? This is, of course, the fundamental dichotomy between basic and applied research. The answer, perhaps, is "both." But then one might ask a more difficult question: Are the insights from basic research ("understanding") useful for applications ("improving")? Does the transfer from theory to practice ever work in education?

2. *Are there any enduring results?* This challenge is repeatedly raised by mathematicians and other traditional scientists. Where are the scientific laws or "theorems" of this field? Have we learned anything that was not already known to good teachers everywhere? Are there any results from educational research that are as credible as the best of traditional scientific research? Has the science of education advanced at all since Plato? Is educational research merely a rediscovery of known truths—or worse yet, a camouflaging of wisdom in obscure language?

3. *Is educational research significant?* One test of significance is replicability. Indeed, the credibility of science rests on the ability to reproduce results under identical conditions. In education, one wonders if the variables can ever be controlled sufficiently to meet this stringent test. If no two situations are ever identical, how can one be sure that any result is reliable? What are the characteristics of reliable educational research? Must replicability be one of them? How can research be vested with meaning and significance in the absence of replicability?

4. *Is educational research predictive?* The ability to predict events is another test of scientific research, especially valuable in fields such as astronomy and geology where controlled, replicable experiments are rare. Where are the results of educational research that predict results? More challenging, are there any surprising predictions? Is educational research any better than intuition as a predictive tool?

5. *What does educational research explain?* Some parts of science (e.g., string theory) are judged not so much by their empirical results as by their contributions to grand theories that help explain how things fit together. Much of mathematics fits this paradigm: the deepest results are those with greatest explanatory value. So one must ask of educational research: what does it explain that we didn't know beforehand? Does it teach us anything new? Are there any aha! insights?

6. *Is educational research about teaching or learning?* Do the insights from what is learned about teaching have any connection with student learning? Do insights about how students learn actually help teachers improve their teaching? More fundamentally, can educational research verify any meaningful relationships between teaching and learning?

7. *Is educational research useful?* Is it believable, reliable, persuasive? Is the body of educational research sufficiently strong to convince skeptics? Can the results of research, if properly communicated, change people's instincts and beliefs? We know that university faculty—mathematicians especially—do not include educational research among the repertoire of relevant and important knowledge required of a practicing university or college faculty member. But even if they had this knowledge, the question remains: is the evidence powerful enough to change habits of mind?

8. *Does it work?* This is, after all, the bottom line. The purpose of education is to educate, so it is not unfair to ask of educational research whether the students of those who read and understand all that is written about educational research learn more mathematics than those who study with mathematicians who disdain educational research. To answer this may require educational research on the effects of educational research. Who would believe that?

9. *What findings of educational research are relevant to student learning of mathematics?* What has been learned about cognition that explains how students learn mathematics? What has been learned about social factors that influence the way students learn mathematics? Are there cognitive, developmental, or social factors in learning that are unique to mathematics?

10. *Is education susceptible to research?* Is education more an art than a science? The many examples of students who succeed despite "bad" teaching and students who fail despite "good" teaching fuel suspicion that education is less a science than an art, where beauty is as much in the eye of the beholder as in the mind of the creator. Can this suspicion be addressed by evidence? Can research help determine just what part of education is really scientific?

11. *What are the right questions to ask?* Much educational research is primarily a sophisticated analysis of particular students' learning under particular conditions. Other research probes the mind of individuals in order to learn how students learn. Should educational research inquire not only into whether students learn, and how, but also into what they should learn? Should it inquire into what teachers need to know in order to lead students to learn? (Do teachers who know more mathematics make better mathematics teachers? Do Ph.D.'s make better teachers of undergraduate mathematics?)

12. *Is statistical theory the right tool for educational research?* Students are, after all, not as predictable as plants growing in variously treated patches. Why is a methodology of analysis and significance testing that was developed as a tool for agricultural research the most widely used tool for educational research? Is it even possible to carry out double blind experiments in education? Might newer methods (ethnographic, philosophical, historical) provide better insights? Are statistical methods in education provably more replicable, more predictive, or more explanatory than other methods?

13. *Is it possible to fairly assess educational innovation?* A common question for educational research is the question of comparison: is treatment A (calculus with calculators) better than treatment B (calculus with chalk)? But what is the basis for comparison? Does one measure both groups on their use of both chalk and calculators? Just how can one fairly compare groups educated in classes with different objectives? Without fair comparison, is it possible to assess innovation? How can we determine whether the innovation will survive transplantation to a new environment?

14. *Dare we do educational research on college students?* Undergraduate courses represent the last chance for students to learn mathematics. How can faculty justify subjecting such students to unproven methods for the sake of research? Will students and parents tolerate having their taxes and tuition used to turn courses into experiments and students into guinea pigs?

15. *Does the act of research change the outcomes of research?* The reputation of the "Hawthorne" effect has dogged educational research much as the uncertainty principle has limited the horizons of quantum physics. Since the efforts put forth by students are known to be influenced by the degree to which someone pays attention to them, the very act of research will almost surely introduce a significant confounding factor in observed results. As important as this technical difficulty may be, perhaps even more important is the degree to which widespread belief in the Hawthorne effect, whether warranted or not, undermines credibility of research no matter how carefully done.

16. *Can one hear the signal amid the noise?* The linkage between teaching and learning is mediated by numerous factors whose variability is enormous and largely beyond the control of any researcher. In this environment, the observed effects of the few variables we can control are quite likely to be indistinguishable from the many we cannot control. Can we be sure that effects we observe are due to the causes we have created?

17. *Who is qualified to conduct research in undergraduate mathematics education?*
    There are specialists in educational research, and specialists in mathematics
    research. But what qualifies one as a specialist in mathematics education
    research—particularly at advanced levels where the nature of mathematical
    practice becomes more dominant? Is it necessary to have advanced degrees
    both in education and in mathematics? Can one be credible in both worlds
    without such credentials?

18. *Are the results publishable?* More to the point, will the research still make
    valuable reading ten or twenty years later? Might reports of practice, no
    matter how carefully (e.g. statistically) documented, be more like "news"
    than like "research"? Is educational research primarily of ephemeral value,
    useful for the moment but not for the ages? If so, does that diminish its value
    for the moment?

19. *Will anyone read it?* How can the mathematical community persuade mathe-
    maticians to read and learn from published research reports? Research on un-
    dergraduate mathematics education is a scattered literature, with no flagship
    journal and no editorial traditions. Research will be read only if it becomes
    de rigueur, if lack of knowledge of major papers yields professional embarrass-
    ment. For that to happen, one must answer a prior question: will one miss
    anything of vital importance by remaining ignorant of educational research?

20. *Does it count for tenure?* At last, the only question that really counts.

St. Olaf College

RESEARCH IN COLLEGIATE MATHEMATICS EDUCATION

EDITORIAL POLICY

The papers published in these *Research in Collegiate Mathematics Education* volumes will serve both pure and applied purposes, contributing to the field of research in undergraduate mathematics education and informing the direct improvement of undergraduate mathematics instruction. The dual purposes imply dual but overlapping audiences and articles will vary in their relationship to these purposes. The best papers, however, will interest both audiences and serve both purposes.

CONTENT.

We invite papers reporting on research that addresses any and all aspects of undergraduate mathematics education. Research may focus on learning within particular mathematical domains. It may be concerned with more general cognitive processes such as problem solving, skill acquisition, conceptual development, mathematical creativity, cognitive styles, etc. Research reports may deal with issues associated with variations in teaching methods, classroom or laboratory contexts, or discourse patterns. More broadly, research may be concerned with institutional arrangements intended to support learning and teaching, e.g. curriculum design, assessment practices, or strategies for faculty development.

METHOD.

We expect and encourage a broad spectrum of research methods ranging from traditional statistically-oriented studies of populations, or even surveys, to close studies of individuals, both short and long term. Empirical studies may well be supplemented by historical, ethnographic, or theoretical analyses focusing directly on the educational matter at hand. Theoretical analyses may illuminate or otherwise organize empirically based work by the author or that of others, or perhaps give specific direction to future work. In all cases, we expect that published work will acknowledge and build upon that of others – not necessarily to agree with or accept others' work, but to take that work into account as part of the process of building the integrated body of reliable knowledge, perspective and method that constitutes the field of research in undergraduate mathematics education.

REVIEW PROCEDURES.

All papers, including invited submissions, will be evaluated by a minimum of three referees, one of whom will be a Volume editor. Papers will be judged on the basis of their originality, intellectual quality, readability by a diverse audience, and the extent to which they serve the pure and applied purposes identified earlier.

SUBMISSIONS.

Papers of any reasonable length will be considered, but the likelihood of acceptance will be smaller for very large manuscripts.

Five copies of each manuscript should be submitted. Manuscripts should be typed double-space, with bibliographies done in the style used by American Mathematical Society and Mathematical Association of America journals. Manuscripts should eventually be prepared using either AMS-TeX (amsppt) or AMS-LaTeX (amsart). The macro packages are available through email without charge.

CORRESPONDENCE.

Manuscripts and editorial correspondence should be sent to one of the three editors:

Ed Dubinsky
Department of Mathematics
Purdue University
West Lafayette, IN 47907
(bbf@sage.cc.purdue.edu)

Jim Kaput
Department of Mathematics
University of Massachusetts, Dartmouth
North Dartmouth, MA 02747
(JKAPUT@umassd.edu)

Alan Schoenfeld
School of Education
University of California
Berkeley, CA 94720
(alans@violet.berkeley.edu)